P9-DIZ-345

The DEER HUNTER'S Almanac

The DEER HUNTER'S Almanac

Edited by Sid Evans

THE ATLANTIC MONTHLY PRESS
NEW YORK

A SPORTS AFIELD BOOK

Published simultaneously in Canada.
Printed in the United States of America.

Library of Congress Cataloging-in-Publication Data
The deer hunter's almanac / edited by Sid Evans
and the editors of Sports afield.
p. cm.
These pieces originally appeared in the Sports afield magazine and Sports afield almanac.
ISBN 0-87113-643-0
1. Deer hunting. I. Evans, Sid. II. Sports afield.
SK301.D3814 1996
799.2'77357—dc20 96-18944

Design by
Michael Lawton

Atlantic Monthly Press
841 Broadway
New York, NY 10003

1 3 5 7 9 10 8 6 4 2

Contents

As the buck bounded down the mountain with a goodbye wave of his snowy flag, I realized that he and I were actors in an allegory. Dust to dust, stone age to stone age, but always the eternal chase!

—Aldo Leopold, *A Sand County Almanac*, 1949.

Deer hunters after a three-day hunt in Wisconsin, 1880.

A taste of venison after a productive hunt, 1897.

A Vermont hunter with a sizable buck around the turn of the century.

Introduction

IF YOU THINK ABOUT HOW long human beings have been hunting deer, you might wonder why we aren't any better at it. Deer eat the same things, mate during the same season, fight and rut and bed down in the same kinds of places. They leave an embarrassing amount of sign wherever they go—tracks, rubs, scrapes, piles of pellets. They cannot see things in color, and in many places, they are the biggest animals in the woods. We hunters should have no problem finding them. Some of our friends who do not hunt are amused when we come home empty-handed, weekend after weekend. Everything

about hunting deer looks easy to them—you sit in a stand, wait for a deer to come along, then shoot it. Where is the excitement in that? What is the point?

It is probably not worth explaining to these people that it's not the dead deer we're really after. It's not pride or vainglory and it's not even the wonderful-tasting meat, though these things are certainly more translatable once we've returned to civilization. At bottom, it has more to do with a remarkable feeling that hits us when we are watching a deer moving slowly through the woods, and we are alone. There is no mystery in a white-tailed deer standing at the side of the road, dumbfounded by the cars racing by at 65 miles per hour. There is no mystery, either, in a 10-point trophy buck mounted on the wall. It's the remembrance of something else—that remarkable feeling—that makes the trophy important. It's a commemoration of a rare moment, where a hunter made the effort to seek out a deer, and was fortunate enough to glimpse one in its own environment, on its own terms. It's the fact that despite everything we know about deer and their very predictable habits, we really know nothing. In a sense, we are all amateurs.

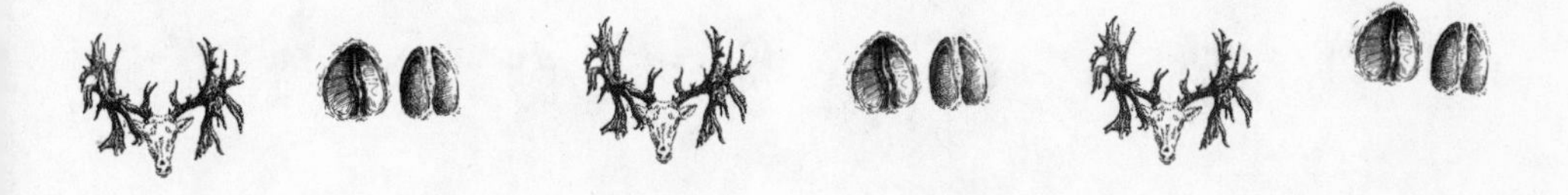

There is not a successful deer hunter in the world who has not come up with his own peculiar methods—some of them secret, some not—for beating the long odds of killing a deer. The Ojibwa Indians of the Great Lakes figured out they could attract deer by smoking wild aster in a pipe, the smell of which was like the scent of a deer's hooves. Other tribes—such as the Choctaws and Cherokees in the Southeast—would carry skinned-out deer heads on their belts, which they could wear over their heads whenever they needed to make a stalk (this is no longer an advisable, or legal, technique). They used decoys and calls, and they knew that banging a pair of antlers together could summon a buck during the rut. They hunted with bows and arrows, and before that, were fond of blowguns that shot poison darts. They regarded deer with such reverence that many of their myths and ceremonies glorified the animal, and for some tribes a young man's passage into manhood could be consummated only when he had taken his first deer, and returned to camp with the proof.

Even in an age of high-powered rifles and GPS units, our reverence for the deer has not changed. Maybe it is because a 250-pound white-tailed buck can disappear as quickly and adroitly as

any animal in North America—some Native American tribes believed that it actually did disappear. Or maybe it is because an entire herd of does and fawns can move through the woods so quietly on a foggy December morning they seem more like ghosts than anything real enough to eat. Only a hunter can understand that kind of magic. It simply doesn't matter what kind of gear we're equipped with or how powerful the scope is on our 7mm rifle; we are still human beings and will never be able to see, hear or smell as well as the dumbest deer in the woods. For us, deer are so compelling precisely because they can fool us almost every time. Almost.

Those of us who hunt deer are not the kind of people who tend to give up easily. Over the years we have learned a thing or two, and it seems that every time we go into the woods for another hunt we learn something else. Some of this helps us to take a deer and actually bring it home, but we probably learn a lot more from the hundreds or thousands of deer we don't shoot, the ones that pass by us without ever knowing we were there. Even when we don't learn anything, it always makes us feel lucky just to see them.

In this book we have tried to compile some of the best information and most interesting pieces written about deer in *Sports Afield* since the magazine was founded in 1887. There were not as many deer to hunt back then, but over the last quarter century deer populations have boomed in nearly every state but Alaska and Hawaii, and so have the articles written about them. Many of these pieces originally appeared in the *Sports Afield* Almanac, which was introduced by Editor Ted Kesting in 1972; others appeared as departments or short features. All told, more than 250 deer hunters contributed, making this, we hope, a very unique look at what is now America's favorite game animal. Some of the contributors—like Dwight Schuh and Peter Fiduccia, Tom McIntyre and Ted Kerasote—are what we would call pros. They have hunted, studied and written about deer all their lives. Others are just guys who wanted to share a couple of their best deer-hunting secrets. Do not be surprised if you turn up some contradictory views. There's more than one way to shoot, skin and cook a deer; but it may be that the best way of all is the one you have to figure out on your own.—***S.E.***

1
Deer and Their Habits

"Thump, thump, thump, thumpity-thump!" I do not know how better to describe the sound that came down wind to me—that indescribable sound a deer makes when running leisurely. My heart kept time to his hoof beats.

—O. Warren Smith, *Sports Afield*, December 1916.

Pre-Rut Action

The most overlooked stage, and the one least capitalized on by hunters, is the pre-rut. Also called the false rut, this typically occurs in early October. Archers who hunt during this time are privy to a phenomenon where, all of a sudden, within a 24-hour period, they begin to find a multitude of fresh scrapes. What causes this obviously intense breeding change in bucks? The onset of the pre-rut.

Because there is a lack of does in estrus in early September, bachelor herds of bucks continue with their normal behavior until early October, when a brief estrus cycle (18 to 36 hours) occurs. This brief cycle is brought on by the mature whitetail does (4 ½- to 5 ½-year-olds) that are coming into their first estrus period for the year. Once bucks discover the estrus pheromone, their natural reaction is to start making scrapes and rubs. They'll continue to do this for up to 36 hours, all the while dashing throughout the woods in search of does.

If you learn to recognize the new scrape and deer activity signs as part of the pre-rut period, you will dramatically increase your chances of shooting a buck.

Once bucks discover the estrus pheromone, they're ready to scra

By setting up around areas where bucks have opened new scrapes, for example, you can simply wait them out, figuring that sooner or later a buck will come to check out his scrapes in hopes of finding a doe in estrus. Or you might try attracting a passing buck with a deer call or rattling antlers. Another good tactic during this period is to make a mock scrape and agitate a buck into responding. All of these tactics will work during the pre-rut as bucks are eager to respond to the first signals of the start of the breeding season.

It is important to keep in mind that the pre-rut does not last long. Generally, throughout the East, the pre-rut occurs about the 7th through the 10th of October, give or take a couple of days either way.

Since only a few bucks actually get to mate during the pre-rut, the rest of the bucks become quite frustrated. This frustration is nature's way of laying the groundwork of behavior and activity that precedes the primary rut. Every doe experiences an estrus cycle every 28 days until she is successfully bred. Therefore, a hunter can count 28 days forward from when he first discovers the above-mentioned activity during the pre-rut and he will place himself squarely within the primary rut.

The Peak of the Rut

The primary rut begins about the last week of October and continues until about the last week of November, depending on what part of the country you live in. Think of that time frame as a graph containing peaks and valleys.

Bucks are most aggressive during the primary rut.

There are several obvious signs after pre-rut that indicate the start of the primary rut. Bucks are seldom seen traveling together. They have an ever-increasing intolerance for one another and will often engage in immediate aggressive behavior upon simply encountering other bucks. In addition, because of the building tension and continuing decrease in daylight, they will take out their aggression more often on saplings or trees. This activity also helps get them physically prepared for the inevitable battles that will come over the next few weeks.

Unlike the pre-rut, when only a few does come into estrus, the primary rut has a majority of does coming into their estrus cycle. This accounts for a dramatic increase in doe activity, too. Hunters will see does checking out buck scrapes, making scrapes of their own, running erratically through the woods, and avoiding fawns. Like female dogs, as they come closer to actually accepting males, they will begin to flag their tails. For deer, this simply means they will be seen carrying their tails in a horizontal position, off to one side. Hunters who take advantage of all of the above signs will find themselves at the right place at the right time.

Buck Fever

If you think human teen-agers do some odd things, consider the poor male deer, which goes through puberty year after year. As hunters know, a buck in rut may do some strange things.

Post-Rut Action

Count 28 days from the primary-rut date, and you will have the prime time of the post-rut, also called the late rut. At this time, most of the immature (or latest-born) does, and any other doe that has yet to be successfully bred, will come into estrus. Many hunters have had success with rattling during the post-rut. In many parts of the country, rattling can be effective into January and even February, as some does are still experiencing estrus cycles.

The post-rut usually begins after a distinct low period in the primary rut. Throughout the East, the start of the post-rut usually begins around mid-December and runs for about 28 days. It is a period that is overlooked by many hunters who are sure that the rut is long over.

One sign to look for during this phase is a quick and dramatic increase in deer activity. Does that come into estrus during this period seem to have an intensity about finding bucks, and will trot along, depositing estrus urine. They will frequently make soft grunts, too. Any bucks scenting the estrus pheromone will quickly swarm to these estrus does. Hunters will sometimes see several bucks chasing one doe. It's almost as if the bucks have come to realize it's their last chance.

Long Live the Deer

Deer have been known to live for more than 15 years, but an 8-year-old in the wild is rare.

Logging More Deer

Deer tend to congregate in areas where logging of hardwoods has recently taken place. They love to munch on the downed treetop branches.

Deer frequent these areas particularly in late autumn and early winter, a time when summer food is about gone and they are changing their diets in preparation for winter yarding. This time coincides with open deer seasons in most northern states.

To take advantage of this situation, check with loggers and find out where they are working. Preseason scout these areas for fresh treetops and branches. Construct blinds or, if legal, plan to use tree stands for early-morning and late-evening hunts.

What You Should Know About Deer Glands

White-tailed deer have several external glands that play a significant part in their communication and behavior. These glands include the orbital, tarsal, interdigital, metatarsal and preorbital. Pheromones created by these glands are received and interpreted by deer. These olfactory messages act to alert, calm, attract, frighten, identify and even assist in establishing a deer's rank within the herd.

✳ ***Orbital:*** One gland that both does and bucks use for marking during the rut is found on the forehead. Officially known as the orbital gland, this gland—a scent marker that specifically identifies the deer who made it—is used when a deer rubs against trees, saplings and twigs. After rubbing and depositing this forehead scent, bucks and does will smell and lick at the rubbed areas, which, weather permitting, will carry the odor for several days.

✳ ***Tarsal:*** The tarsal gland is located on the inside of the hind legs of all deer. This gland turns almost black as bucks continually urinate on it throughout

When Deer Eat Cedar

The next time you are hunting near a cedar swamp, notice how far you can see into it. The visibility is good because deer have browsed on all the cedar leaves for as high as they could reach. Though deer like cedar greens, they do not usually feed on them until there is at least six inches of snow on the ground. Then their more preferred browse is covered.

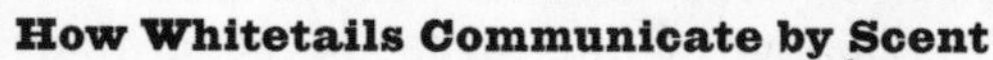

How Whitetails Communicate by Scent

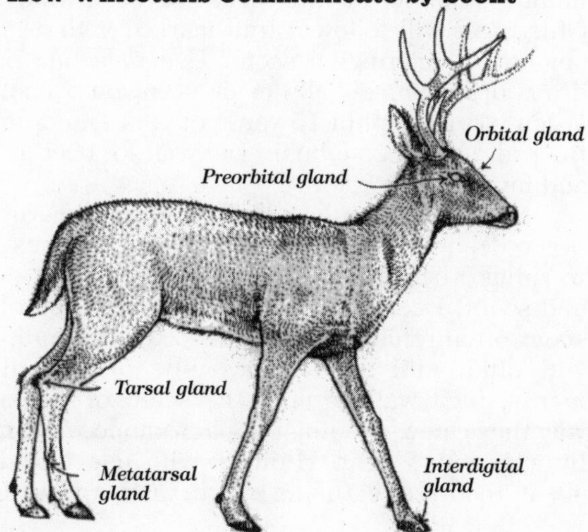

the rut. Deer use tarsal gland pheromones several ways: as a visual and olfactory signal of a mature buck; as an alarm; to identify individual deer; and, in mature deer, it becomes involved with breeding behavior during the rut. When the deer becomes excited, the hairs on the tarsal gland stand erect. All deer urinate on these glands, which contributes to their odor.

To obtain optimum response from tarsal scent during the rut, put several drops of commercially made scent in a drip-boot dispenser, rag or pad and hang it from a branch near your location. Its scent will permeate the area and act as an attracting, or agitating, smell for deer. Don't place it directly on your clothing, as you don't want deer zeroing in on you; instead place the scent 10 to 20 yards away from your position.

Some biologists believe the metatarsal gland, at the back of the legs, emits a pheromone.

✳ ***Interdigital:*** The interdigital gland is found between the deer's toes. This gland emits a yellow, waxy substance with an offensive, potent odor. Interdigital scent is like a human fingerprint. It is individual to each deer.

Hunters can use interdigital scent in two ways. Used sparingly, it acts as an attractant. All deer leave minute amounts of interdigital scent as they walk. Other deer will follow a trail marked with a normal amount of interdigital scent. Use only one or two drops of a commercially made scent on a boot pad. When you are within 15 yards of your stand, remove the pad, hang it on a bush, and wait for deer to come and investigate.

Deer also use interdigital scent as a warning to other deer. When a deer stomps its hooves, it is warning other deer of danger through sight, sound and scent. Deer that come upon excess interdigital scent often refuse to continue down the trail. They will often mill about nervously for several moments, then walk around the scent or retreat the way they came, heeding the pheromone warning left by a previous deer. Hunters who use interdigital scent incorrectly (using more than a few drops) may inadvertently spook rather than attract deer.

✱ ***Preorbital:*** The preorbital gland is located just below the inner corner of the eye. Its main function is to act as a tear duct. However, deer will continually rub the gland on bushes, branches and tree limbs, especially during the rut. Biologists speculate the gland is used to deposit a specific pheromone to mark certain areas and to help deer identify one another.

One white-tailed buck identifies another by its orbital gland.

✱ ***Metatarsal:*** This gland lies within a white tuft of hair located on the outside of the hind legs just above the dewclaws. Some biologists believe the gland is atrophying, and therefore has no purpose. Others say the gland emits a pheromone, and is used by deer for identification. In any event, this gland is not well understood. When you're using metatarsal scent, be ready for anything to happen. You may be fortunate and attract deer, or unfortunate and spook them. Give it a try where you hunt and see what happens.

My, What Big Eyes You Have

Deer are extremely well equipped to see under dim light conditions. A deer's eye has a great number of light receptors known as rods. In comparison, human eyes have fewer rods and a greater number of cones, which are color receptors. Rods are far more sensitive to light than cones.

In the eye of a deer and other animals with good low-light vision, there is a layer of reflective pigment. The layer is called the tapetum, and it increases the night vision of deer. If light passes through the eye of a deer without sufficiently stimulating the rods, the tapetum bounces it back, providing another chance for the light receptors to respond. It is reflection from the tapetum that shines when a bright light strikes a deer's eye.

While a deer's eye surpasses ours in its ability to gather light, its vision does have shortcomings. A deer cannot see well at long distances, nor can it perceive colors as accurately as can humans.

Weather-Related Deer Movement

Researchers monitoring radio-collared deer have found that their movements increase greatly just before and after thunderstorms, and that deer activity is slowest during hot weather. More important to the deer hunter: Activity increases in the fall, peaking during the rut.

Identifying Whitetail Foods

Most hunters can identify agricultural crops. But when it comes to shrubs, weeds and other deer foods, it's a different matter. Following is a brief guide to many of their favorites:

Both red- and white-oak acorns are common whitetail browse.

✳ ***White Oak Acorns:*** Deer will feed on almost any acorns, but if those of the white oak are abundant, whitetails will feed on them and ignore everything else, even corn. Identification of the oaks is fairly simple. White oak leaves have rounded lobes, while those of the red oaks and black oaks have pointed lobes. The acorns mature in one season and are sweet, not bitter. Acorns from the red and black oaks mature in two years and are bitter.

Identifying Deer Foods

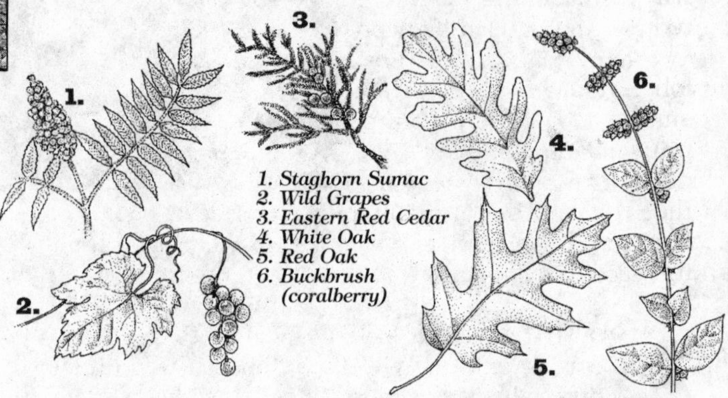

✳ ***Buckbrush:*** Also known as coralberry, this shrub covers most of the eastern half of the United States, and is a prime wintertime food of whitetails. It is easily distinguished by the dark red berries that last throughout the winter and into the early spring.

✳ ***Wild Grapes:*** In winter and summer, look for a number of different varieties of wild grapes, but don't confuse them with moonseed. Grape leaves are coarsely toothed. Moonseed leaves and berries resemble grapes, but the leaves are smooth and untoothed.

✳ ***Sumac:*** The sumacs include a wide variety of plants, and even some of their names bespeak deer. Staghorn sumac is found throughout most of the whitetail's range and is easily identified by the upright clusters of bright red berries. Smooth sumac, fragrant sumac and dwarf sumac all are favorites of whitetails.

✳ ***Eastern Red Cedar:*** When times get tough,

whitetails will eat the berries and boughs of these trees. With their dark green needles and blue berries, they are easy to identify.

Deer Food

The word "maple" might bring "syrup" to mind. But maple is also a mainstay in the diet of white-tailed deer. Two favorites of these browsing animals are red maple and striped maple, or moosewood. Buds, leaves and tender branches of both are among the whitetail's favorite forest foods. Both grow throughout hardwood forests, with the maple variety being more widespread than the striped species. Locate either in woodlands, and you're in deer territory.

You're probably more familiar with red maple, whose buds, flowers, leaf veins and leafstalks have reddish tints.

Striped maples are dwarf trees with leaves of three lobes that turn brilliant yellow in fall. The bark is smooth, thin and greenish with white stripes, not unlike the striped pattern of zebras.

Deer browse on other kinds of foliage, too, but these maples rate high on their "preferred" list. Hunting clubs often increase the deer-carrying capacity of ranges by planting and propagating these two maple trees.

Stripped of Their Buckhood

Through some untimely misfortune, a male deer can occasionally become a neutered recluse—an impotent animal who can no longer reproduce the species or carry a respectable rack of antlers. Their antlers, if any, will emerge as a deformed mess of mismatched tines and abnormal conformations.

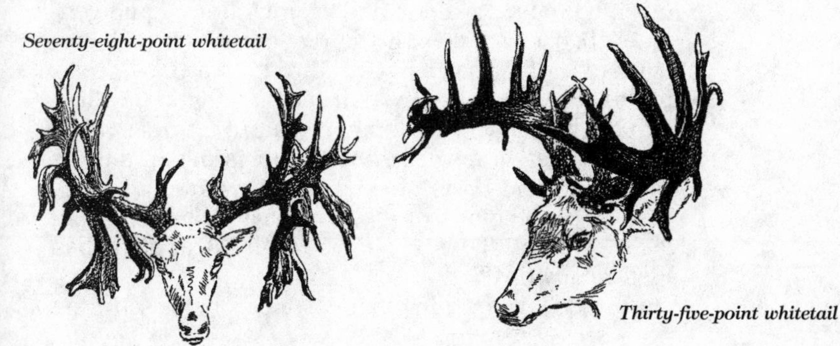

Seventy-eight-point whitetail

Thirty-five-point whitetail

A white-tailed buck—antlers shed—scratches the resulting itch.

Antler growth is initiated by the pituitary gland. Then as the immature mass of velvet-encased blood vessels reaches its maximum size, an increase in testosterone seasons each antler and stops its growth. This usually occurs in September. The blood supply is then cut off, the entire formation hardens and the velvet case dries up. The buck becomes a dangerous fighting machine, ready for battle. When the rut is over, testosterone levels begin to decline, the antlers are shed and the cycle is complete. However, if that same buck were to become castrated, the lack of hormones would prevent the antlers from developing and maturing and keep the velvet from shedding.

A doe can also produce antlers. Her antlerless skull is not due to lack of male hormones but to certain female hormones that prevent their growth. If that female hormone is not present in necessary amounts, antlers will occur.

Herd Life and Attrition

A mule deer herd is not a constant. Animals interact with each other in different ways depending upon the time of year and the cycle of life they are in, and during any given time, not every deer in the herd responds in the same manner. For example, during fall most mule deer bucks are in bachelor herds or are living singly and apart from the group. However, it's not impossible to find a mature buck with an autumn herd of yearlings, fawns and does. Bucks fight during the rut, but not all bucks.

Mule deer tend to be the most active in the early morning, late afternoon and evening hours, and generally bed down during the day, although in harsh winter weather they may feed throughout the daylight hours. Sometimes they will feed at night. Feeding habits change with the time of year, too, from springtime grazing on new grasses to subsisting on shrubs in the fall and browsefodder in the winter. In some areas, they may also feed in cultivated fields, especially if these fields exist where the deer herd has historically dwelled.

The behavior of bucks and does changes during

the rut. Although the mating season varies in different parts of the country, it usually begins in late October and continues into December, sometimes even into February. The bucks that were traveling together, or even singly, in the summer and early fall pick up interest in the ladies by late fall. While mule deer bucks do not gather harems, as pronghorn will, they do try to cover a variety of does, taking advantage of almost any opportunity.

Does bear their young in summer, with June a prime month for birthing. The gestation period is 210 days. An early hunting season, by the way, does not thwart the mating cycle. A buck-to-doe ratio of 1:13 is more than sufficient to service the does.

As long as the species has a place to live, a 40 percent attrition rate maintains herd size; normal fawn survival increases the herd by 40 percent annually. Of course, the 40 percent figure includes all attrition, not just hunting harvest. A single mountain lion, for example, can feast on 50 to 60 deer in a year. Coyotes take fawns. I know of at least one incident in which a mature doe was killed by coyotes. Disease may also take a toll. When herd numbers are overly large, disease is especially deadly because of contact between animals of different herds. Bad winters can be worse than all other herd-reducing factors combined, especially where critical winter range has been taken over by human development and the deer cannot retreat to their usual hard winter feeding grounds. Reduction of habitat is the No. 1 threat to the mule deer, as it is for

The Trees a Deer Will Eat

Deer have a hard time finding enough to eat when there is deep snow covering their browse all winter. Consequently, kindhearted people will put out hay for them—but deer cannot eat hay. They are generally browsing, not grazing, animals. When there are hay drops in our national parks, they are usually for elk, which do graze. Winter deer browse on pine, hemlock, mountain maple and swamp maple, and will chew up twigs as thick as a pencil.

A 40 percent attrition rate is necessary to maintain herd size.

other wildlife. The number of deer in the West will continue to thrive, offering a very important annual hunter harvest.

How Teeth Show a Deer's Age

Hunters often judge deer to be much older than they are. This is particularly true where the buck has attained good body weight and carries an attractive pair of antlers.

A new premolar with two cusps indicates a deer 1 ½ years old.

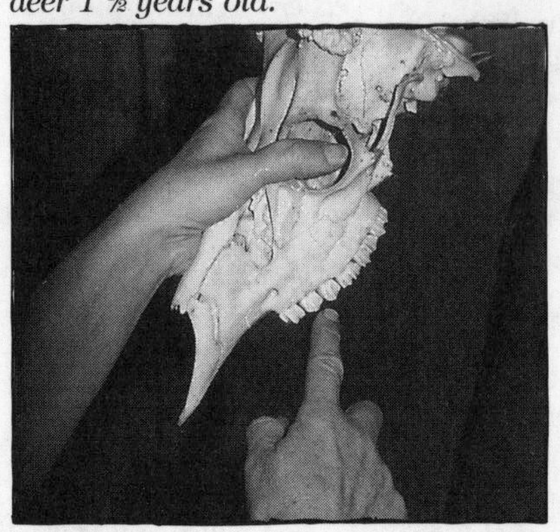

A careful look at the third premolar tooth should tell the story. At 1 ½ years the emerging third premolar is usually pushing upward on the milk tooth it is replacing. This third premolar tooth will have three points (cusps) as a milk tooth; the new one has two cusps. Sometimes the new tooth has already pushed away the older tooth, but its enamel will appear newer and whiter than that of the other back teeth, with no stains from chewing. This, too, clearly indicates that the deer is very likely near 1 ½ years old.

To better inspect the teeth, you must cut or scrape the gum away from the jaw to expose the root surface and alveolar border of the jaw.

Facts About Antlers

The sight of a whitetail or a mulie sporting a trophy rack always sets your blood racing and your heart thumping, but most hunters don't understand the chemistry that created those antlers.

Antler growth begins in early spring and is triggered by increasing sunlight. Blood vessels run beneath the soft velvet skin of the new antlers and carry minerals for their growth. Blood running under the velvet helps cool the deer in summer.

It takes about four months for antlers to reach full size. A bull moose rack can reach 45 pounds or more. Biologists rate moose antlers as the fastest-growing tissue known.

Deer antler growth depends on a buck's heredity, but diet plays a big hand, too. Nearly all whitetail bucks will sport forked antlers by their second winter if their diet consists of 12 to 18 percent protein and large amounts of minerals. More trophy racks are produced in areas where whitetails can dine on high-protein soybeans.

Spikes Can Hurt the Herd

There are different schools of thought about a buck whose antlers are spikes. If it is genetically inferior, and if it fathers fawns, the herd may suffer. Unless injury or nutritional deficiency prevent good antler growth, the spike buck is unlikely to ever be much of a specimen. Therefore, culling such animals from the herd is considered good herd management.

Nature makes sure that a yearling buck's body growth takes precedence over antler growth. After three to four years, as the body matures and growth slows, he begins to carry trophy antlers. The biggest racks appear in a buck's prime, at about five years of age.

A whitetail sheds his antlers anytime from late November into March. The timing usually depends on the success of the mating season, because shedding is also connected to the level of his hormone testosterone and the number of unbred does in his territory.

The "New" Mule Deer

Mule deer country varies enormously. Mulies inhabit high forests, piñon-juniper stands, aspen patches, oakbrush thickets, sagebrush flats, riverbottoms—anyplace that has food, water and cover. The West is a mosaic of all of these.

There is great satisfaction in getting a mulie buck one-on-one, but there is equal satisfaction in pulling off a drive that nets a good buck or two. The finesse

Have mule deer gotten smarter with increased hunting pressure?

of such a drive takes as much thought and preparation as a stalk.

Mule deer were once thought of as dummies, and rightfully so. But they have become a changed animal and are now fully as crafty as an elk or a whitetail. As hunting pressure increased, beginning in the 1960s, the "new" breed became shiftier and better able to avoid hunters. Hunter-success ratios dropped from close to 90 percent to as low as 35 percent in some states, with a lot of the credit going to the mule deer itself. This new mulie mentality has made driving a very popular method.

Mule deer once migrated in the fall and spring, some as far as 60 miles. These days there may be no migration at all or a move as short as five miles, from deep-snow high country down to the valleys. Mule deer learn a lot more about the country they live in because they don't roam as far, and they are more likely to use trails than ever before. Big bucks have learned to live closer to human habitation. Cultivated crops attract the deer, and they have learned to hide in relatively populated areas.

Most mule deer driving is done silently, except for the occasional use of a whistle. Sound carries a long distance in the West, so shouting and crashing through the brush is unnecessary. The silent approach is more likely to confuse the deer and make him go where you want him to.

When a strong wind is blowing, deer don't move around much. If you hunt under these conditions, find hollows, thickets and coulees sheltered from the gusts and still-hunt slowly, searching for bedded animals.

Moose and Deer Don't Mix

The National Wildlife Federation reports that what's good for the moose in Maine isn't necessarily good for the deer. The state's moose population has experienced a slight boom, from the mere 3700 that existed in 1951 to over 15,000 now. The revival of the moose is seen as a result of the regrowth of pole timber. The white-tailed deer, however, "has been hurt by the regrowth past the stage where they can browse on the hardwoods, and they're really taking a beating from the cutting of Maine cedar swamps."

Well-Fed Deer

An experiment at Penn State University has found a direct correlation between deer size, antler formation and diet. Several years ago, 25 deer were captured as fawns and fed on a rich diet of grass, deer pellets and proper vitamins and minerals. Some 15 ½ months later their weights and antlers were compared with those of deer killed by hunters on what was considered poor range. Average live weight of the 10 study-deer was about 140 pounds. Deer from free range averaged only 108 pounds. Differences in antler growth were even greater. Poor-range bucks could grow only spikes. The well-fed study deer of the same age averaged almost six points apiece.

Deer Injuries

A study of 1002 injured deer collected throughout the Southeast by researchers and biologists with the Southeastern Cooperative Wildlife Disease Study yielded the unexpected information that only 2.3 percent of the injuries resulted from gunshots or arrows.

The study also reported that 76 of the examined deer, or about 7.6 percent, had been injured once before. The percentage of injured deer did not differ according to sex or physical condition of the animal, but injuries were more likely among older deer.

Hunting pressure did not influence the likelihood of deer becoming injured. Gunshot wounds and highway collisions with automobiles were usually fatal.

What a Deer Needs

You can get a good idea of where exactly to hunt deer (or other game) based on the animal's needs: sun, shade, food, a place to rest, a place to hide, water, dry ground, a windbreak. Consider the weather, time of day, hunting pressure and other factors. Then head for the appropriate place. For example, if it's cold and the ground is mostly snow-covered, deer would probably need sun (warmth) and food. Consider hunting a sunlit south slope, where the ground is more likely to have some bare spots and possibly acorns.

Whitetail Facts

Most of the time, deer will smell you before they see you.

✳ ***Antlers:*** All males of the deer family grow a new set of antlers each summer. Antlers can grow as much as half an inch a day. They are soft and, if damaged, grow irregularly. Once fully grown, the antlers harden, and the deer rubs off the velvet covering on trees and brush. These buck rubs establish dominance among competing males. Diet is the primary factor affecting antler size, so you cannot age a deer simply by the number of points on its antlers. Pennsylvania biologists found that the greatest antler growth seems to take place between a buck's third and fifth years. In winter all males shed their antlers, which are then eaten by rodents (they are a good source of calcium).

✳ ***Communication:*** Deer snort, grunt, and whistle; fawns bleat. When alarmed, deer flash their seven- to 11-inch-long whitetails to signal danger. As a deer scrapes the ground in search of food, it leaves an odor from the scent glands between its toes, which others can easily detect. Urine is also used as a means of communication.

An alarmed buck will often grunt or snort to alert others.

✳ ***Description:*** Hoofed animals, white-tailed deer are four to six feet long and 38 to 40 inches high at the shoulder. Their hair is reddish in summer, brownish gray in winter. Record body weights for bucks have reached over 400 pounds.

✳ ***Diet:*** Strictly vegetarian, deer browse and graze on twigs, leaves, bark, buds, nuts, grass, grain, herbs, mushrooms, vegetables and fruit. They also grub for roots.

✳ ***Habitat:*** Deer are found in forest edges, swamp borders, woodland openings and suburban backyards.

✳ ***Habits:*** Active in the early morning and at dusk, white-tailed deer form small family groups led by old does. When the moon is full, deer are usually more active at night.

Bucks keep antlers strong by "horning" in the mud.

Moist Antlers Are Strongest

Biologists have discovered that dry deer antlers are much more brittle than moist ones. Bucks prevent their antlers from drying out by "horning." They thrash their antlers in leaves, and push them into mud. Besides providing moisture, this also helps build up a layer of protective resin to hold it in. Were it not for horning, antlers would be very fragile and delicate.

✳ ***Home Range:*** The average home range of a whitetail in the East is about one square mile and overlaps the ranges of other deer. They use different parts of the area, day and night.

✳ ***Hunting:*** Most state deer hunts are conducted in late autumn at the end of the rut, when bucks are most active. In Pennsylvania, 45 percent of the deer killed during gun season are shot before 10:00 a.m. on the first day. Hunters' culling of bucks and older does creates a stable population.

✳ ***Life Span:*** Where hunted, most whitetail bucks live only 1 ½ years, though the natural life span for deer in the wild can be 11 to 12 years. Captive deer have survived for nearly 20 years.

✳ ***Locomotion:*** A white-tailed deer can run about 35 mph and sprint at least 45 mph. They gallop, trot and leap, and can jump 8 ½ feet vertically and 27 feet horizontally. They are good swimmers.

✳ ***Mortality:*** More than 350,000 deer are killed annually in car accidents across North America (46,000 in Pennsylvania alone). In addition to roadkills and hunting, deer die from winter starvation, disease and parasites, and from large predators.

✳ ***Population:*** There are more whitetails inhabiting North America today—over 20 million—than there have been during any other time in recorded history.

✳ ***Range:*** Thirty-eight known subspecies of white-tailed deer can be found from Canada to northern South America.

✳ ***Reproduction:*** A doe gives birth to one to three fawns, depending on her nutrition. The pregnancy rate is 90 to 95 percent in most whitetail populations. Six deer in a Michigan enclosure—two bucks, four does—produced a population of 222 in just seven years.

✳ ***Senses:*** Deer have very good senses of hearing,

sight and smell, but smell is the most acute. They may be able to detect the odor of danger from half a mile away. Though deer do not see colors as humans do, their eyes can pick up the slightest movement, even the blinking of a hunter's eye.

✳ ***Winter Survival:*** A deer's heavy winter coat of hollow hairs (twice as many as in summer) insulates it against the cold, and a 100-pound deer can be sustained on a mere half-pound of digestible nutrients a day. Even so, many deer die when snow is deep and food inaccessible.

Why Wide Spreads?

Every season a buck sheds his antlers, the length of the pedicles shortens. The reduction is not even, however. The outer part shortens more than the inner, and that is why a buck's antler beams have a greater spread each year.

Wheezing, Snorting, Stomping, Grunting

Deer use eight distinct sounds to communicate with each other. Researchers and biologists from Mississippi State University recorded all sounds made by individuals in a captive herd of 40 deer over a one-year period. Each sound was then analyzed and labeled:

✳ ***Plain Grunt:*** Sound made by males during rutting season and also by females when searching for young deer.

✳ ***Bleat:*** Call made by fawns and other young deer when seeking attention, food or security.

✳ ***Nursing Whine:*** Made by fawns during feeding.

✳ ***Snort-Wheeze and Aggressive Snort:*** Both used during hierarchical and territorial disputes among adults. The snort-wheeze is the less intense.

✳ ***Distress Call:*** Made by all deer in times of extreme travail, such as when attacked or caught by predators.

✳ ***Hoof-Stomp and Alert-Snort:*** Both are warnings. When a deer first detects a predator, it seeks to warn others in the herd by stomping. This sound is low-level, so it's not likely to be heard by the predator. If the danger becomes critical, the deer will sound one loud alert-snort for all to hear, then sneak away to safer ground.

Where Bucks Bed Down

Deer living in flat terrain use thick, brushy or swampy areas for daytime bedding; in mountainous habitat, bucks often bed high on knolls and in hard-to-reach craggy areas. Hunt near these early and late in the day to catch the animals going to or coming from feed sites.

The Perils of Adopting Deer

People occasionally adopt deer fawns under the mistaken impression that they are orphans. But they are very likely not orphans. A fawn that you find alone in the woods is almost never alone; usually the mother is hiding nearby, waiting for you to clear out so that she can claim her kin.

Most adopted fawns die in captivity.

Breaking up a family of deer isn't the only harm you'll be doing by adopting a fawn. It takes special feed and milk formulas to raise a healthy deer and few people know just what those formulas are. Even if you raise a fawn into a healthy adult, it will be an inquisitive nuisance if it's a doe and downright ornery if it's a buck.

Perhaps the best reason not to adopt a deer lies in one motherly duty that most of us aren't willing to undertake. During the first three or four weeks of life, fawns must be stimulated in order to defecate. As the fawn is nursing, the doe licks and nuzzles her fawn's anus for a minute or so, and only then will the young deer relieve itself. Most young fawns that die in captivity do so because their need for help in defecating is neither recognized nor fulfilled.

Whitetails vs. Mule Deer

The white-tailed deer is one of evolution's outstanding success stories. *Odocoileus virginianus* has inhabited North America for some 3 million years, descending from ancient deer species that migrated from the Old World. Today 38 species of whitetails can be found from southern Canada to northern South America, and at more than 20 million, the whitetail population in this country is at not merely a historic but very possibly a prehistoric high. When it comes to areas where whitetails and mule

Typical Antlers

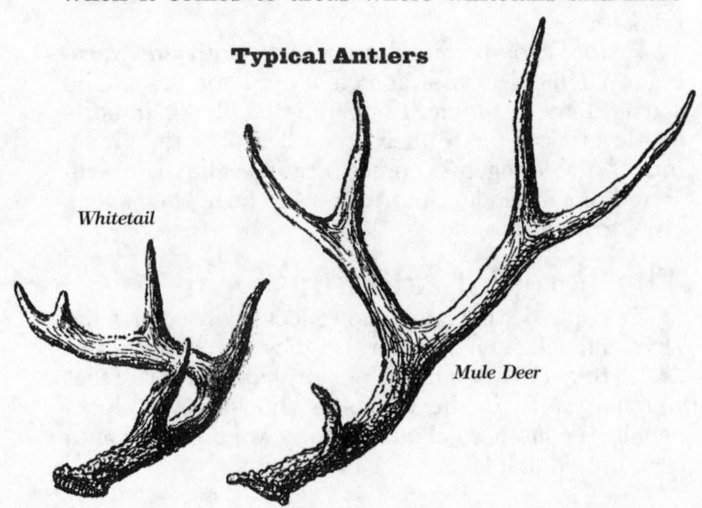

Right-from-the-Start Shooting

No other sport compares with shooting in its prompt and unforgetful reward of a right start in it.

The boy who learns to shoot right immediately begins to see his possibilities, to realize his tremendous advantage.

Recognition of the importance of *Right*-from-the-Start shooting is the foundation of Remington UMC service.

Start your boy off right too — there is no premium to pay and much for him to gain.

Our Service Department will introduce him to *Right*-from-the-Start shooting, pass him along to the National Rifle Association steered clear forever of the handicap of bad shooting habits and qualified to try for the *official decoration for Junior Marksman.*

This is the only *official* decoration of its kind. It is authorized by the U. S. Government.

As he learns the value of right methods, we believe he likewise will learn to appreciate right equipment, and join the many thousands who prefer Remington UMC.

Boys—Write at once for the Four Free Remington *Right*-from-the-Start Booklets on Shooting, and mention this advertisement.

THE REMINGTON ARMS UNION METALLIC CARTRIDGE COMPANY, INC.

Largest Manufacturers of Firearms and Ammunition in the World

WOOLWORTH BUILDING NEW YORK

deer overlap, though, whitetails may be too much of a good thing.

As environment sciences professor Dr. Valerius Geist has pointed out, where the ranges of the two deer cross, the whitetails' more successful defense behavior and ability to breed with mule deer does—while whitetail does will not mate with mule deer bucks—result in a net increase in whitetails and a decrease in mule deer. As whitetail ranges in the West expand, this fact should be addressed by wildlife departments.

Where the Whitetails Are

Wildlife managers are reluctant to estimate deer populations in their area for an obvious reason: You can't count deer as easily as you can cattle. Despite that, we are obsessed with numbers, and wildlife administrators in most states satisfy that desire by providing us with their best educated estimates. When it comes to white-tailed deer, there's no contest for first place. Texas has more than 4 million whitetails.

After that, in the 1.5 million class are two states: Michigan and Alabama. One million, plus or minus: Wisconsin, Georgia, Pennsylvania and Mississippi. Three-quarters of a million: New York, Virginia and Florida. Probably in the next 10 for numbers (but for which current figures "aren't available") are Louisiana, South Carolina, Minnesota, Missouri and West Virginia. Kill figures are more accurate than population figures.

Hunting on the Beaver Moon

Deer hunters must consider a staggering number of variables—from wind and weather conditions to terrain and food supply. One factor that influences deer all over the country is the presence of the full moon. Deer have a strong tendency to feed and move about in the moonlight and to sit tight during the day. The result is that large numbers of tracks can be seen come morning but the deer themselves tend to be scarce. So, when the moon is full, hunt near likely daytime bedding areas.

Indians called the full moon the Beaver Moon because of the flat-tailed rodent's propensity to work on a bright night. Take a tip from the Indians and alter your deer hunting accordingly.

2

Scouting and Tracking

I have been told that a deer with a heart wound will invariably run as long as it can hold its breath. The old-time hunters held to this theory, as to a belief that their rifles "shot level" and that black bears were crazed and blinded by the August sun.

—Roger Reed, *Sports Afield*, March 1901.

Where to Find Early Season Deer

An often successful alternative to hunting trails and food plots is hunting feeding areas, possibly early acorn trees. In the South, pin oak and white oak acorns are among the first to fall, and both are heavily used by deer. Big white-oak trees, which may grow by themselves or with only a few other scattered oaks around them, can draw deer like magnets and are certainly worth watching.

Scouting trips before the season will help you locate such trees, as will past experience. Deer will use the same trees season after season. Again, the key is getting your stand into a position that offers a clear, short-range shot.

Other foods may also be eaten during the early autumn bow seasons. Depending on which part of the country you live in, they may include a wide variety of grasses, vines, briars and fruits.

Old farms are among the prime places. You may locate some of these by driving country roads, or you

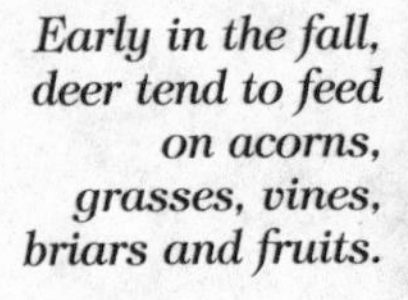

Early in the fall, deer tend to feed on acorns, grasses, vines, briars and fruits.

can pinpoint them on a topographic map. Try to find a secluded farmhouse that has been abandoned for years and forgotten. Deer hide in the tall grasses that grow up around the old buildings; they may eat the fruit from old orchards and drink water from a still-flowing spring.

Carefully scout old homesteads to locate specific trails, bedding and feeding areas and other activity zones that could be observed from a tree stand. By studying the overall picture of the farmhouse and the surrounding cover, you can sometimes put a complete movement pattern together. Then put up a tree stand where you can intercept a deer moving from one activity area to another.

Ten Tips for Better Scouting

1. Start early. Late summer or early autumn. Some hunters begin scouting the day after deer season closes.

2. Learn your hunting area. To plan effectively, you need to know the locations of ponds, thickets and different foods, and where trails and ridges lead.

3. Don't spend time looking for deer. Look instead for sign—it's faster, easier and more productive. Seeing deer is a bonus.

4. Don't over-scout. Too much human scent and presence in an area will eventually cause bucks to change their movement patterns.

5. Scout on foot, not by vehicle. You'll see and learn more.

6. Scout different places. Pay especially close attention to water crossings, grain fields, hardwood forests and pine thickets. Deer will utilize many areas during a full season.

7. Don't key on summer food sources. In autumn, food choices will probably change.

8. Continue to scout even as you hunt. Especially in states with long seasons, deer activities change, and a hunter must be aware of these changes.

9. Try to understand the overall picture. Recognize deer activity in your area by looking for several different types of sign. For example, don't plan your entire hunting strategy around the presence of tracks. Look also for droppings, signs of feeding and scrapes and rubs.

10. Scout for fresh sign. Don't concern yourself with old sign; also, look for repeated activity, not a single occurrence.

How to Scout Whitetails With Statistics

Remember all those big-game statistics released by your state conservation agency—the ones you barely glanced at? They can be a valuable aid in bagging a whitetail.

Most states record the deer harvest by county or region. They also hold trophy-measuring programs. A careful examination of these records can tell you where the largest harvest is occurring and where the trophy bucks are being taken.

Regions that tend to produce trophies also tend to produce larger-than-average bucks overall. Compare the total harvest figures of the different areas you hunt. You may find that an area that has a high success rate with trophies also presents you

with consistent opportunities to take a sizable buck.

When you study big-game harvest figures, compare several years of data (a sudden increase in the kill may be due to a one-time special hunt). Look for an area that reflects a steady or increasing harvest over several years.

Get Back

There are too many stories about wounded big game returning to where they were first hit to ignore it as an occasional occurrence. Perhaps because deer and elk are herd animals and feel safe in numbers, the wounded animal often comes back to the location where it was first shot to find its companions. Whatever the reason, if you fail to recover the animal, go back and search the area where it was shot.

Estimating Deer Density

Studies done in recent years by Wisconsin's Department of Natural Resources have demonstrated that white-tailed deer populations can be accurately estimated by simple deer trail observations. And although these studies were done for purposes of deer management on a large scale, the same methods can be easily utilized by individual hunters to scout hunting areas. Having a rough idea of deer density, which can vary from two to 25 deer per

Whitetail density varies roughly from two to 25 deer per square mile.

square mile, can greatly increase a hunter's chances of bagging a buck by helping him choose good hunting sites.

Before defining the method, here are a few useful facts to know while scouting for deer:

Trails that deer are likely to be on during a fall hunting season are established in the late summer and early fall. The trails become visible as plant growth halts and are complete by leaf-fall. The adage of scouting for deer after the first black (killing) frost of autumn holds true, but a hunter can begin his survey much earlier if he wishes. Habitat type or soil compaction do not affect observability of deer trails. A hunter should not worry about missing so-called "hard to see" deer trails. These either do not exist, or represent one-time passage of a few deer.

Learn to Use Hunting Pressure

If you have no alternative but to hunt heavily used public hunting areas, don't despair. Early in the season, learn to use the hunting pressure there to your advantage. The large influx of hunters from easily accessed points pushes game—be it deer, ringnecks or rabbits—into small areas. Find these areas, or strips of cover leading to them, and you will be successful.

A simple survey method will provide a hunter with a good estimate of deer density. It can be done alone if the hunter is familiar with the area he is in. Otherwise, a companion is recommended. In either case, a compass and topo map are needed.

First find a clearly defined tote road as a jump-off point. Place any type of marker on the road, then enter the woods. Follow a straight course, perpendicular to the road, and walk at a comfortable pace. Count all the trails as you cross them. A logging road, streamside fisherman's path or backpack trail may be included if they show abundant deer signs.

After about 450 yards into the woods, you have completed your first "run." Turn right, go straight for 200 yards, then start your second run, this time back to the road. Count the trails again as you cross them. Once back on the road, move down another 200 yards and do another pair of runs. Do as many runs as possible in the time you have.

You arrive at the estimate of deer density in the area by finding an index number. Divide the total number of runs you made into the grand total of trails you crossed. If you counted 24 trails and made eight runs, for example, your deer-population index number is three.

Index numbers of two, three and four can be roughly translated to a deer density of three to seven deer per square mile. Index numbers of five, six and seven mean populations of between seven and 13 deer per square mile. Beyond these, more trails per run simply mean more deer.

Reading Deer Sign Language

How to interpret and take advantage of the rubs, tracks, droppings, holding areas and scrapes that whitetails leave behind:

✳ ***Rubs:*** One of the most evident forms of deer sign is the buck rub. Bucks rub their antlers from late August until they shed them, often as late as March. The rub phenomenon is 10 times greater in frequency, size and intensity in the buck's range from October to December. Rubs found during this time indicate where the buck is traveling to seek out does. Fresh rubs found from January on indicate the location of a buck's core area—an ideal place to plan an opening day archery ambush.

Buck rubs will also show you a predictable pattern regarding where a deer is traveling, what time of day he's using that particular route and what food source he heads for most frequently. Most important, rubs can betray a buck's bedding area. For instance, when a hunter locates a spot where a buck has rubbed as many as 18 or more trees within 50 yards, he has dis-

Rubs can be used to determine the size and travel patterns of a buck.

covered the buck's core area—where a buck spends 90 percent of his time. Bucks will rise several times a day from their beds to change position. Often, as they stretch, they will walk a few feet, defecate, and rub brush or a sapling before lying down again. By paying attention to wind direction and locating the most heavily used trail into this area, a hunter can plan a successful ambush.

The buck rubs most hunters come across can also be used to determine the travel pattern of a mature white-tailed buck throughout his entire home range. When you discover a rub, lie down or kneel with the rub in front of you. Slowly look left and right. You should discover other rubs on trees, saplings or brush in front of you. Usually, each additional rub will be about 30 to 50 yards ahead of the other. Also, by knowing whether you are heading to or from feeding or bedding areas and which direction the rub is facing, you can tell if he is using this route in the morning or evening.

When Deer Feed

Deer have four major feeding times: daybreak, noon, dusk and midnight. Don't neglect hunting during midday, when good deer are often on the move.

There are some misconceptions about rubs, too. A seasoned hunter never allows himself to be fooled into believing that all large rubs are made by large-racked bucks. This just isn't so. Many small bucks, spikes and forkhorns rub large saplings (six to 10 inches around) totally bare of bark as high as four feet from the ground. Don't be duped into thinking the rub was made by a buck with a large rack. Generally, small-racked bucks rub large trees more often than most hunters believe.

Large-racked bucks, however, seldom rub smaller trees because of the lack of resistance the tree offers them. Trophy-sized bucks spend much less time rubbing their antlers on trees than they do thrashing and tearing up thick, resistant brush. When you find a large bush with roots that have been dislodged and branches broken and strewn about, you have found an area where a trophy buck has displayed his aggressiveness for all subordinate deer to see and smell.

Rubs made on branched trees 16 to 20 inches wide showing bark stripped on both trunks have most likely been made by bucks with wide racks. Rubs with many deep gouges are usually made from burrs and kickers on the antlers of mature bucks. If the tree or sapling is dripping sap, the rub was probably made within 48 hours. The most exciting aspect of finding a rub is that you will know it was made by a buck in your hunting area. If there are fresh shavings hanging from the tree and on the ground, look around—he may be only 100 yards ahead of you, establishing another rub marker.

✳ ***Tracks:*** Tracks provide evidence that deer are living and traveling in specific areas. The clues left by tracks can often lead to more confusion than information, however. For instance, finding a single deep track that shows the dewclaws only means you have discovered the track of a heavy deer—buck or doe. Beyond that, and despite what many old-timers might say, most biologists agree that it is impossible to definitely identify the sex of the deer from its tracks.

There are some dependable indications left by tracks, however, which will help a hunter make a more educated determination as to the probable sex of the deer. For example, tracks that show the spreading of the toes on a hoof and do not appear to be pigeon-toed generally belong to a buck. If you are following tracks that meander throughout the woods, you can bet you are on the trail of a doe. Bucks walk with a purpose. Their tracks will often move from point A to point B while taking the path of least resistance. When a buck meets an obstacle, he will often walk around it (unlike a doe, which will often walk under it) and then resume his line of direction. In addition, when a buck urinates, he drips urine into his tracks. A doe squats and urinates in one place.

Seeing tracks with drag marks behind them is almost as good as actually seeing the buck himself standing in the tracks. If you know what you are looking for, you can tell a great deal about the size and state of mind of the animal you're following.

Bucks are built differently at the hips than does, and sway and rock slightly side-to-side when they walk. This gives a buck the tendency to drag

Diagonal Walkers

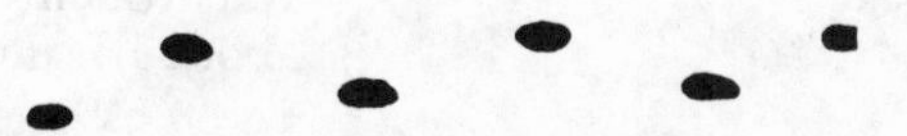

Note: *Each circle contains both front and rear feet. Typical of the deer, dog and cat families. The direction of travel is from left to right.*

Bounders

Note: *Large circles are rear feet. Small circles are front feet. Typical of the weasel family.*

Gallopers

Note: *Large circles are rear feet, small circles are front feet. Typical of rodents, rabbits and hares.*

Pacers

Note: *Large circles are rear feet, small circles are front feet. Typical of opossums, raccoons, bears and members of the weasel and rodent families that have wide bodies.*

his hooves, especially in the snow, rather than daintily lifting them as a doe will. The deeper and longer the drag mark, the older and heavier the buck. To further confirm your findings, measure the size of the track. Generally, tracks that measure 4 ½ to 5 ½ inches long belong to 2 ½-year-old bucks. You can bet your hunting boots that tracks longer than 5 ½ inches are from mature bucks (3 ½ to 6 ½ years old).

✷ ***Droppings:*** Another good indication that deer regularly use an area is the number of fresh piles of droppings you find. According to the Wisconsin Department of Natural Resources, deer defecate about 13 times during a 24-hour period. Therefore, if

Deer Dropping Formations

you are hunting an area with a high whitetail population, analyzing this sign becomes a little more difficult. Here are some hints to help you learn the information you need. Adult buck droppings are clustered and are larger than adult doe droppings. A good rule of thumb to determine the approximate age of a buck is that a single pellet measuring about ¾ inch is from a buck about 2 ½ to 3 ½ years old. Larger pellets up to 1 ⅜ inches long are usually from the truly trophy-sized animals. Mature buck droppings are thicker, longer and generally clumped together in a shapeless mass rather than in single pellets. Finding a few dozen such droppings in an area perhaps 50 yards around is an indication you have found a buck's core (or preferred) bedding area. You can ignore the smaller loose pellets usually found in piles of 10 to 50 droppings; these belong to fawns, yearlings and does.

All deer droppings will change in texture and consistency as the deer's food source changes. This information will help you determine whether deer are browsing or grazing, and where you should post accordingly. If a deer is feeding on grasses or fruit, the feces will usually be in a loose mass composed of soft pellets. If you find such sign, hunt along routes leading to alfalfa fields or apple orchards.

When a deer is browsing on drier vegetation, such

as twigs, branch tips and acorns, its droppings will be less moist because of a lack of mucus. Consequently, the pellets do not stick together when expelled from the deer. They will be hard to the touch, longer than usual and more separated. If you find this type of sign, plan your ambush accordingly. Posting in an alfalfa field when most of the deer dung you are finding is hard and long will most likely prove futile, as the deer's droppings suggest it is not grazing but browsing.

When inspecting deer excrement, remember that weather affects its look and texture. If the pile you are examining is on top of a ridge, for instance, and is subject to wind and harsh sunlight and receives no moisture, it will dry much more quickly and appear to be stale, when in fact it may be less than two days old. In places where deer pellets receive shade and moisture and are not exposed to wind, the opposite holds true.

You can avoid the guesswork by picking up a few pellets and squeezing them between your thumb and index finger. Remember that pellets dry from the outside in. So one that looks fresh on the outside but is hard and dry and begins to break apart when squeezed is at least four days old. Shiny, fresh-looking pellets that have the consistency of Play-Doh when squeezed may be very fresh.

You will often discover the freshest sign in and around bedding areas. When you locate an area with eight to 10 beds grouped within a short distance of one another, you have found the bedding area of does and fawns. But it pays to investigate such a site. You can determine if a buck is within the group by measuring the size of the beds. Beds measuring about 2 ½ to three feet long are those of does, yearlings and fawns. However, beds measuring about 3 ½ to 4 ½ feet long are most likely those of mature bucks.

✱ ***Social Areas:*** Deer tend to mill about in these areas before continuing to their destination. I refer to these spots as "social areas." They are often located where there is low brush, foxtails and cattails, and are usually found between bedding and feeding areas. Deer that congregate here leave behind discernible sign. These areas always contain numerous droppings of all sizes and matted high grass where deer have temporarily lain down. There is an excess of heavy trails leading into and out of these spots. And

Indian Tracking Trick

Deer tracks can become difficult to detect and even seem to disappear in certain light conditions. When faced with this frustrating situation, use the old Indian trick of moving from side to side to get the light in the most favorable position. With the light at the right angle, the hoof impressions or disturbed leaves will become apparent because of the shadows they cast.

Move from side to side to get a look at hoofprints.

almost every social area you will ever encounter will be within 100 hundred yards of both bedding and feeding areas.

Social areas are especially good places to ambush deer in the evening. The best way to avoid confusing a social area with a bedding area is to look for available food. Social areas do not provide much in the way of food. Most offer only browse. A few will have overgrown grasses such as timothy. It is simply a semi-protected area for deer to hole up in prior to feeding in the evening or bedding in the morning.

✳ ***Scrapes:*** The most pertinent information about scrapes is that they are sexual calling cards of both bucks and does. Although there are some half-dozen types of scrapes, you'll need to pay attention to only a few. The first scrapes are made in early October, and I refer to these as pre-rut scrapes. These are made when the most mature does of the herd come into a brief estrus cycle, usually lasting for less than 36 hours. Bucks who come across this early-season, unexpected scent respond by making a number of small scrapes, which are usually not re-freshened and are the least likely scrapes to produce buck sightings. For the most part, hunters shouldn't pay much attention to them. The same philosophy applies to scrapes made by does, which often do not produce well for hunters. Just use these scrapes to confirm that there are deer in the area. The scrapes that provide the most sightings of bucks are the primary, secondary and gregarious scrapes. Secondary scrapes are the most prevalent and are usually two to three feet round. Bucks will make between 20 and 30 such scrapes throughout their home range during the rutting season. Because of the sheer numbers of secondary scrapes, you can rest assured they are within

Learn to distinguish primary, secondary and gregarious scrapes.

a few hundred yards of a buck's core area. A hunter positioned over secondary scrapes is almost certain to see bucks—eventually.

When the rutting season reaches its peak, the secondary scrapes diminish in number and the remaining ones are used more frequently. These remaining scrapes become wet, churned up, and are transformed into primary scrapes. These primary scrapes, which are usually twice the size of a secondary scrape because of increased pawing, will undergo the peak of activity this time of year. Don't spend a lot of time around secondary scrapes in the early season, but keep track of the scrape activity; when the secondary scrapes become primary scrapes, hunt them until the rutting period is over.

A gregarious scrape can be either a primary or secondary scrape that is used by several bucks of the same age group. Usually they are found in the same areas as secondary scrapes. These gregarious scrapes are nothing more than overused secondary scrapes. Because they are used by submissive bucks rather than the most aggressive bucks, they receive a lot of attention. They can be identified by their irregular shapes and beat-up look. Hunters watching such scrapes will usually score sooner than if they watch any other type of scrape.

All scrapes are made in a relatively straight line over several hundred yards. Scrapes, like rubs, show up in the same places year after year. By identifying the most active scrapes in an area, hunters can count on bucks making scrapes in the same spot the following year. Even when a buck is not near a certain scrape, do not forsake that exact area next year. Bucks are like big trout. If you catch one out of a certain hole, within a few days another fish will have claimed the territory.

✳ ***Putting It All Together:*** When you are out and looking for deer sign, remember to take a small tape measure with you. Doing so will eliminate guesswork, letting you accurately measure the size of droppings, the height and width of rubs, the size of beds and, later, the inside spread of your buck's rack! They're a useful tool for the hunter who reads sign.

The wise hunter doesn't stop reading sign after his buck is down. Now it's time to dissect and examine the digested and undigested contents of the four compartments of a deer's stomach. The information will

Look Ahead

The next time you have to track a wounded deer, try this: Follow the animal's trail by walking along one side of the tracks. You should move quickly in a slightly bent position. Your vision should be focused eight to 12 feet ahead. This creates a sight plane that allows you to see the deer's tracks along the same angle at which his hooves struck the ground. With a little practice you'll see tracks that you might have otherwise missed by standing directly over them and looking down.

let you know where and when the deer was feeding and what it was eating. The least-digested food is what the deer ate last. The food that's been further digested will be mushier, less identifiable, and will have to be investigated more closely—this is what the deer was feeding on first. By knowing the time of day your deer was killed, you will be able to backtrack through the stomach contents to discover the route the deer traveled. This information can be shared with hunting companions who haven't filled their tags. Or it can be used next season if the same general weather conditions and food patterns exist.

Deer sign allows a hunter to eliminate a majority of the guesswork and to place himself at the right place at the right time. Luck now takes a backseat to knowledge and skill. So, by taking the advice of that old Adirondack guide and learning to correctly "read the water," you, too, can become a better deer hunter.

Scouting Secrets for Bowhunters

Scouting for bowhunting is the same as for gun hunting, except for the time of year. Bow seasons generally open in September or October, when deer activity is not the same as it is during gun season. An early-season hunter ought to be prepared for different feeding and bedding habits.

One of those differences is that early in the fall deer might be using grain patches or food plots, rather than eating acorns in the woods. You can determine this by looking for tracks in fields and studying the tops of the individual blades of grain. If they're chopped off fairly uniformly throughout the field, deer are feeding there regularly.

Unless a deer is fairly near the edge of a field, however, your chances of getting one with an arrow are pretty slim. The best tactic is to set your ambush either within range of where the deer normally first enter the field or farther back in the woods along a trail leading into that field.

By walking and studying the edges of a field, you'll be able to determine where deer are entering most often. Frequently, this will be at a corner or near particularly heavy cover. There will generally be a trail leading back into the woods, and it's along this trail that you want to position yourself.

Tree stands offer the best shooting opportunity because they put the bowhunter above a deer's line of

Aging a Track

An easy way to estimate the age of a track is to make a fresh mark alongside it. Use your boot heel, fingers or even, as some hunters prefer, a carry-along deer hoof. Press down hard and compare the marks. The closer they match in texture, sheen and definition, the fresher the original track.

sight and allow some freedom of movement. Some stands are designed for bowhunters and feature a larger platform, since most archers shoot from a standing position.

Your main concern in trail or field watching from a tree stand will be shooting distance. You'll have to be close in order to get a good shot, so look for a tree that offers not only some concealment but also proximity. Because range is so critical, it's a good idea to climb the tree with your stand and see how things look from above. It's better to find out before the season than to learn the sad truth on opening morning that you're out of range.

Early in the fall, bucks can be found near grain plots, or on trails in the woods that lead to those plots.

You may do better backtracking down that entry trail and putting up your stand in the woods. Once deer near the edge of an open field, they naturally become more cautious, and your chances of being detected are greater. You may find several trails converging into one major trail, too.

Know What You're Tracking

Sometimes hunters fail in their attempts to track wounded deer. Such losses could be eliminated by placing proper shots in the first place. Although many hunters aim for the neck or heart, the lungs are actually the best target area for several reasons. The lung cavity offers the largest target, and even if you miss it you still have a good chance of placing a lethal shot. If the bullet hits a little low, it may strike the heart; if high, the spine becomes a prime target; and if slightly forward, the bullet may hit the shoulder.

When a wounded deer darts away, signs left behind usually indicate where the deer was hit. Bright red blood with air bubbles means you've made a lung shot and the animal won't go far. Dark red blood usually indicates a paunch shot—you may find pieces of intestinal fat. A gut-shot deer will bed down if not pursued; if chased he may run for miles.

If blood smears are found high on limbs or bushes, it's likely the deer was hit high, perhaps in the shoulder. Tracks that indicate the deer is dragging a front leg would confirm a shoulder shot, just as dragging a hind leg would usually indicate a hip shot.

Spring Scouting

Do some deer scouting in late winter and early spring before the leaves come out. Deer sign from the previous fall will be highly visible, and you may also spot enough deer to find out where they're most abundant.

Dwight Schuh's Game-Recovery Commandments

Practice the following game-recovery tips every time you shoot, and you'll rarely, if ever, lose an animal.

1. Initial Response

✳ ***Follow up every shot.*** With a bow, you normally know whether you've hit an animal, but with a firearm, it's not always obvious on quick shots or at long range. Always follow up and check for signs of a hit.

✳ ***Shoot again.*** If possible, shoot again to anchor an animal. Hunters are always telling stories about "dead" deer that ran off.

✳ ***Wait and observe.*** Watch the animal's progress as far as possible, and mentally mark his position at the time of the shot and at the point where you last saw him.

✳ ***Mark your position.*** Make sure you can return to your exact shooting spot, if necessary, to restart your search. Surveyor's tape is good for marking spots where you've been.

✳ ***Go to the animal.*** Find the exact spot where the animal stood and search until you find blood, hair, your arrow or other signs of a hit, or until you're convinced you missed cleanly.

✳ ***Think.*** Sit down to carefully analyze the situation; contain your excitement. Game recovery demands a cool head, not frenetic action.

2. Trailing

✳ ***Wait.*** On a chest-hit animal, wait 30 minutes before trailing. On a paunch hit, wait four hours or, on evening hunts, until the next morning. With imminent rain or snow, however, follow immediately.

✳ ***Stick with the trail.*** If the trail gets sparse, you might be tempted to search ahead, looking for the animal. Don't do it. Stay on the trail.

✳ ***Look close.*** If blood isn't obvious, get down on your hands and knees to look for small specks. Look under leaves and blades of grass; blood not dripping onto the ground may smear the undersides of plants as the animal passes.

✳ ***Search for other signs.*** Blood sign could cease, but you might still be okay. Search closely for crushed grass, broken sticks, scuff marks or smeared mud on rocks or logs.

✳ ***Note the tracks.*** Every animal has distinct tracks. Note the size, freshness and features of your animal's tracks; you'll learn to recognize them instantly. Hooves of wounded deer often splay more than normal.

✳ ***Measure stride.*** On hard ground, you can usually predict track placement by measuring the distance from one track to the next. (Use a stick, arrow or gun barrel as a gauge.)

✳ ***Mark the trail.*** If you lose the trail, you're done, so mark it well. Marking the trail also helps give you a line on the animal. Again, use surveyor's tape (and clean it up when you're done).

✳ ***Trail at night.*** On evening hits, wait until dark and trail with a lantern or flashlight.

3. Searching

✳ ***Comb the area.*** If the trail gives out, this is your last resort. Using a compass to stay on line, search back and forth beyond the end of the trail for at least a half-mile. If you have a friend, line up so you can see each other on each pass; if you're alone, make tight

A Topo Tip

Before you go deer hunting next fall, buy a quadrangle topo map for the area you plan to hunt. Take it out on scouting trips and mark down prominent deer trails, areas with rubs or scrapes, and good bedding and feeding areas. You'll soon begin to see a pattern to the deer's movement, and this will allow you to pinpoint the best spots from which to ambush a buck on the season opener.

passes so you can see the ground you've covered.

✳ ***Use binoculars.*** Inspect brush well out ahead. Binoculars help you penetrate brush visually.

✳ ***Don't quit.*** Determination means more than specific techniques. Refuse to give up until you've exhausted all hope.

4. Points to Remember

✳ Very often, animals stop bleeding just before they go down. A dead-end blood trail could mean the animal is near.

✳ Hard-hit deer will often circle an area before bedding down. They may travel downhill or seek water—but don't count on it.

✳ Deer traveling in a straight line, without bedding, may travel a long way and never die.

A Deer Tracker's Guide

Note the differences between the fore (below left) and hind (below right) hoof of a male white-tailed deer.

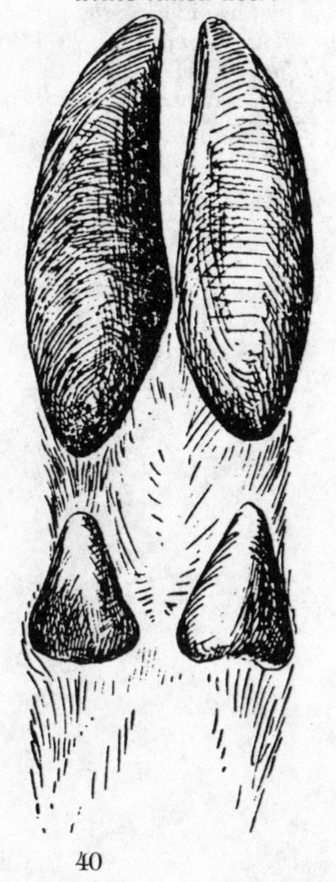

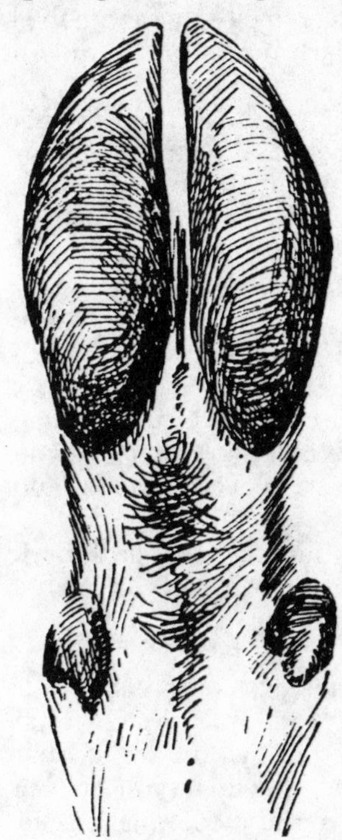

Learning how to read tracks is a three-step process. The first is how to read an individual hoofprint. The next is how to decipher the spacing and arrangement of the imprints. The final step is how to accurately interpret the configuration of the trail.

✳ ***A Word of Caution:*** Reading tracks is like studying grammar. For every rule, there seems to be at least one exception. That said, what follows is generally true, although specific circumstances may lead to exceptions.

✳ ***First Impressions:*** A deer that is walking on soft substrate leaves an imprint in which both halves of the hoof remain close together. A running or bounding deer—indicating that the animal is frightened, alarmed or wounded—leaves an imprint in which the lobes of the hoof are splayed.

Under ideal conditions,

an individual imprint will provide information regarding the age of the track. Generally speaking, a track with sharp, well-defined edges is fresh. Over time, the elements go to work, whittling away at the track until it becomes blurred.

In some areas, winds or other weather conditions will greatly enlarge tracks in a surprisingly short time.

Older bucks will drag their hooves in sandy or muddy terrain.

You can start out following fresh chipmunk tracks and within an hour you would swear you were hot on the heels of a musk ox. The same occurs when the wind blows across powdery snow. On a blustery day, if you come across well-defined tracks in dry sand or powdery snow, you can rest assured that they are only minutes old.

On snowy days, you can roughly gauge freshness by the depth of the snow within the imprint. In rain, it is determined by the depth of water within the imprint and by the sharpness of the track. Rain will round out tracks quickly; heavy rains will make a fresh track look ancient.

Can the sex of a deer be determined by studying a single imprint? Most experts say no. Generally, however, the larger the deer, the larger the track. Further, the heavier the deer, the deeper the imprint. Since bucks are usually larger and heavier than does, the track of a buck will tend to be larger and more deeply imprinted.

A secondary indicator of deer gender is when very small deer tracks are found next to larger ones. This almost always signifies one or more fawns traveling in the company of one or more does.

Some older bucks have a tendency to drag their hooves when they walk. These definitive drag marks are most apparent in shallow sand, mud or snow. This rule of thumb is valid only if the substrate is shallow. In deeper substrate, specifically deep snow, all deer will leave drag marks.

✱ ***In Step:*** Although a buck's stride is generally

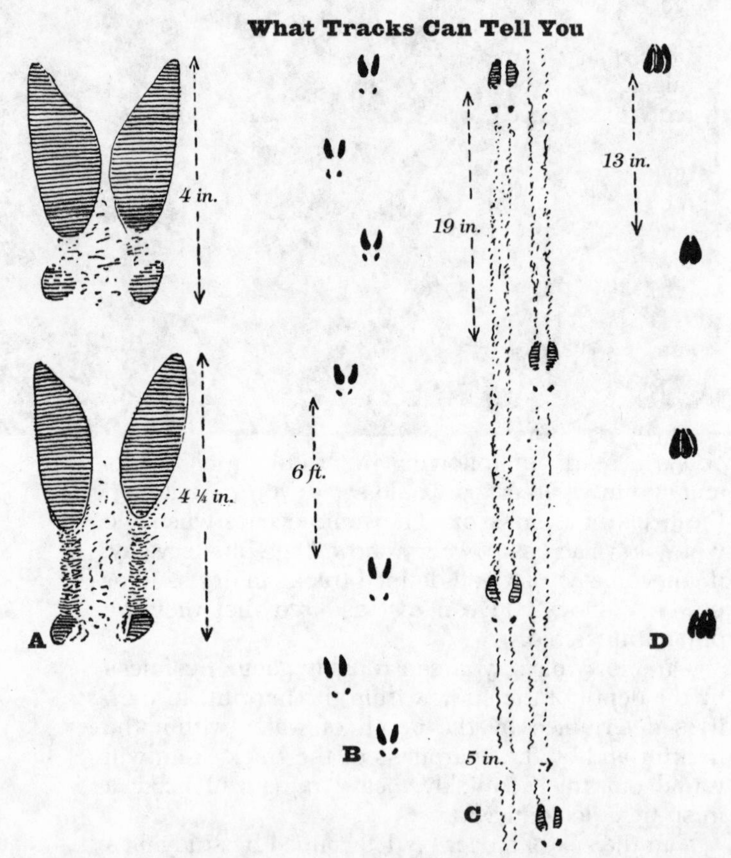

A. Leaping tracks, in mud, with dewclaws showing. In front track (upper) dewclaws are close to hoofs. In hind track (lower) dewclaws are farther from hoof. B. Galloping track pattern, in snow, showing drag marks of toes. C. Walking pattern in snow, showing drag marks of toes. D. Walking pattern of young deer, on dirt road, showing the traditional heart-shaped hoofprint.

longer than a doe's, the spacing between imprints reveals clues to more than just the sex. A walking deer, one that is relaxed and unaware of danger, leaves a trail in which the spacing between the imprints remains fairly consistent, ranging anywhere from one to three feet. What appear to be individual tracks may actually be double imprints, the hind hoof imprinted on top of the front hoof.

A deer that is frightened or wounded will immediately begin running or bounding, leaving a distinctive trail in which the imprints are bunched together in clusters of four. The spacing between track clusters can range anywhere from three feet to 25 feet.

✳ ***Trail Configuration:*** A trail that leads in a straight line, with all imprints facing in the same direction, is indicative of an undisturbed deer traveling at a leisurely pace. One that tends to meander, moving from bush to bush with imprints facing in all directions, is indicative of a feeding deer. A trail that meanders and occasionally describes small circles is indicative of an animal searching for a bedding site.

Of great importance to the hunter is the ability to determine whether or not a deer has been wounded when there is no evidence of blood. Trail configuration can provide two clues. First, a seriously wounded deer will often break into a frantic run without any apparent regard for obstacles. The trail will lead right through thick brush, rock outcroppings and bramble patches. Second, if a deer is traveling with a herd, a wound will frequently cause it to break away from the group and head out on its own. This often occurs immediately in the case of serious wounds that will eventually prove fatal. With lesser wounds, a deer will often remain with the herd for some distance before striking off on its own.

Know What You're Tracking

Veteran deer hunters can usually tell at a glance whether the tracks they're following belong to a buck or a doe. A doe's hind hoofprints fall slightly outside the prints of her front hooves because of a somewhat wider pelvic area (to accommodate birthing).

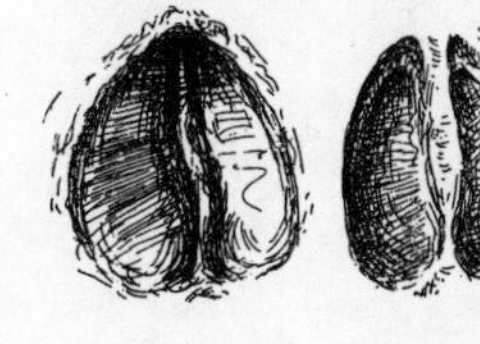

Tracking Sticks

✳ The important thing in trailing any wounded game is not to charge off to where you think the animal will be, but to follow carefully the tracks and blood that the animal leaves.

In the sandy, "fast-healing" soils of southern Texas, tracks can vanish rapidly, so to keep themselves on the line a wounded whitetail is taking, some Texas guides use a tracking stick to mark the trail.

Basically, you measure off the length of the buck's stride on a straight stick and, if you come to a place

Use a straight stick to measure off the length of a buck's stride, and it will make finding the next track much easier.

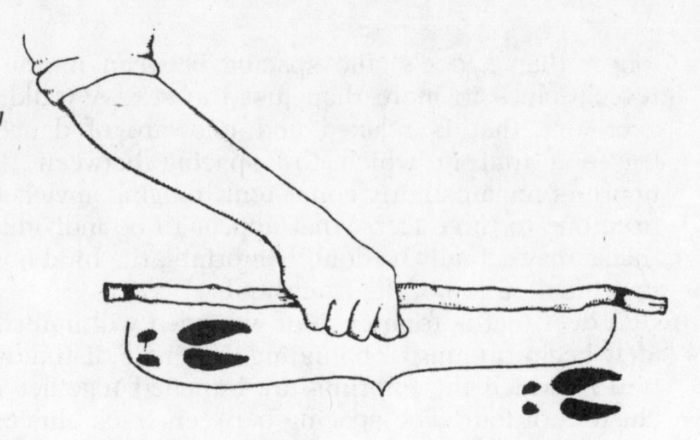

where the trail is indistinct, you can then lay one end of the stick down on the last clear spoor, with the stick pointing in the direction the animal appears to be heading. This will give you a good gauge of where to look for the next track or speck of blood.

If a Deer Is Wounded . . .

In some situations, it is impossible to tell you hit a deer—so unless you know you missed, check the area where the deer was standing when you fired and follow his tracks for at least 100 yards. Look for drops of blood, hair or tracks indicating the deer is stumbling or dragging a leg.

If you find sign indicative of a hit, wait for at least 30 minutes before taking up the trail. A wounded deer that is not pushed, and especially one that hasn't seen, heard or smelled you, is unlikely to travel more than a couple of hundred yards before bedding down. After 30 minutes to an hour, you'll have an excellent chance of approaching within gun range before he gets up—if he's able to get up at all.

First, try working tracks and blood. Those first few tracks, made as the deer bounces away after the shot, probably will be easy to spot. Follow those tracks carefully, looking for blood. Even a small drop here and there will confirm that you're tracking the right deer.

When you can find no more blood, go back to the last drop, or the last set of tracks ahead of it that you know was made by the deer, and resume the search on your hands and knees. Carefully examine the forest floor leaf by leaf. This will probably reveal an occasional blood drop that you missed while walking. Leave a small piece of white Kleenex or toilet tissue near each drop to note the last-known position of the deer and its best-known direction of movement.

If you're unable to find more blood, return to the last-known position of the deer and begin cutting a wide arc through the woods in the last-known direction the deer was headed. Keep a sharp lookout for the deer as well as for more blood sign. Be sure to check out all patches of heavy cover, for a wounded deer is most likely to bed down there. Use your nose and ears. You can sometimes smell a deer several yards away—or hear the buzzing of flies or yellow jackets near a dead deer.

Tracking Is in the Details

The key to good tracking is a patient and disciplined focus on details. Also required is the inductive ability to assemble these details meaningfully—a result mostly of practice.

The initial tracker's task is to identify the kind of print under observation—a fairly easy task in most circumstances, made easier if your attention is fixed on a particular animal. Complete novices can master basic track identification by using a quality field guide.

Next comes the need to age a track—for there's little point in following sign made days or weeks ago by an animal now in another county.

Tracks in wet snow are easier to read because they retain detail.

First, look carefully at the track, which when freshly cut into a soft surface will have a definite and sharp outer edge. The fresher the track, the finer and sharper this rim or edge. The longer the track remains exposed to the elements, the blunter it will become. Frost glazes it; rain or dew flattens it. The inside of the hoofmark, the pad, will normally be clean and flat when fresh; but accumulated bits of grain and dust, or a dry dullness, indicate that the track has been there long enough to weather.

Just how long is a matter requiring some induction. A track made in damp morning earth may dry within hours if exposed to bright, hot sun; but drying may take more than a day if the same track is made in a shaded, cool substrate.

Reading snow takes more refinement, because of variations in the way the stuff falls and settles. Wet snow is usually the easiest for tracking because it retains detail well. A fresh track in wet snow will have a distinct, thin and fine-edged wall between the cloven toes. An older track will show signs of either freezing or thawing, or both. Frozen tracks have glazed-over pads, and the internal ridge lines are hard and icy to the touch. A melting track, conversely, widens out, losing its definition.

Three-Step Scouting

Consider preseason scouting a three-part endeavor. The first step is choosing a good hunting area. The second is searching for sign and learning the lay of the land through footwork and map study. The third is analyzing what you've discovered and using it to pattern the animal's movements and choose the best stand locations.

Powder snow can be tough to read because it's light, dry and easily changed by wind, temperature or physical disturbance. Fresh tracks will have string-tails of crystals kicked up by lifting hooves. When fresh, these crystals are rough and angular; but within an hour or so they glaze and diminish with evaporation. Inside a fresh powder track the rims and center walls are sharp and fine, but tender, easily moved or broken by the slightest touch of a finger. Tracks more than an hour old, depending on wind and temperature, will begin to crust on the inner edges; the harder and deeper the crust, the older the track—except in thawing conditions, in which case the rims become proportionately flatter and blunter as they age.

The wise tracker studies track placement to learn the emotional state of the animal he's following. He knows that track patterns change as a deer walks, trots, runs or gallops. Generally speaking, walking, meandering fresh tracks mean the animal is not aware of your presence; but if they change suddenly into a bunched, wide-splayed, deep-set configuration, you

may have gotten too close or careless, in which case it might be time to find a new set of tracks.

Hunting From the Kitchen Table

Successful mule deer hunters scout thoroughly before opening day, but this isn't always possible. As long as you have a kitchen table, a detailed topographic map and a notepad, however, you can conduct successful scouting missions long before you enter the woods.

A good topo map from the U.S. Geological Survey shows every spring creek, clearcut, campsite and usable road in the area. Timbered land is light green, untimbered land is white. In the West, many access roads that are closed to vehicles remain open to horses, hikers and mountain bikes.

Topo maps also show elevation contour lines, which indicate where to find flats (widely spaced lines); valleys (widely spaced lines with creeks); and steep slopes (tightly packed lines). Once you have learned to interpret these contour lines, the map will take on a three-dimensional look.

It is now a simple matter to mark likely feeding sites and locate nearby bedding areas. For example, most mule deer like to bed in tall, cool old growth on north-facing slopes in August and September, and prefer south-facing slopes in colder weather.

Scout on Time

Do the bulk of your deer scouting two or three weeks before the season. If you scout much earlier, foods, the stage of the rut and deer movement patterns may change before opening day, invalidating your findings.

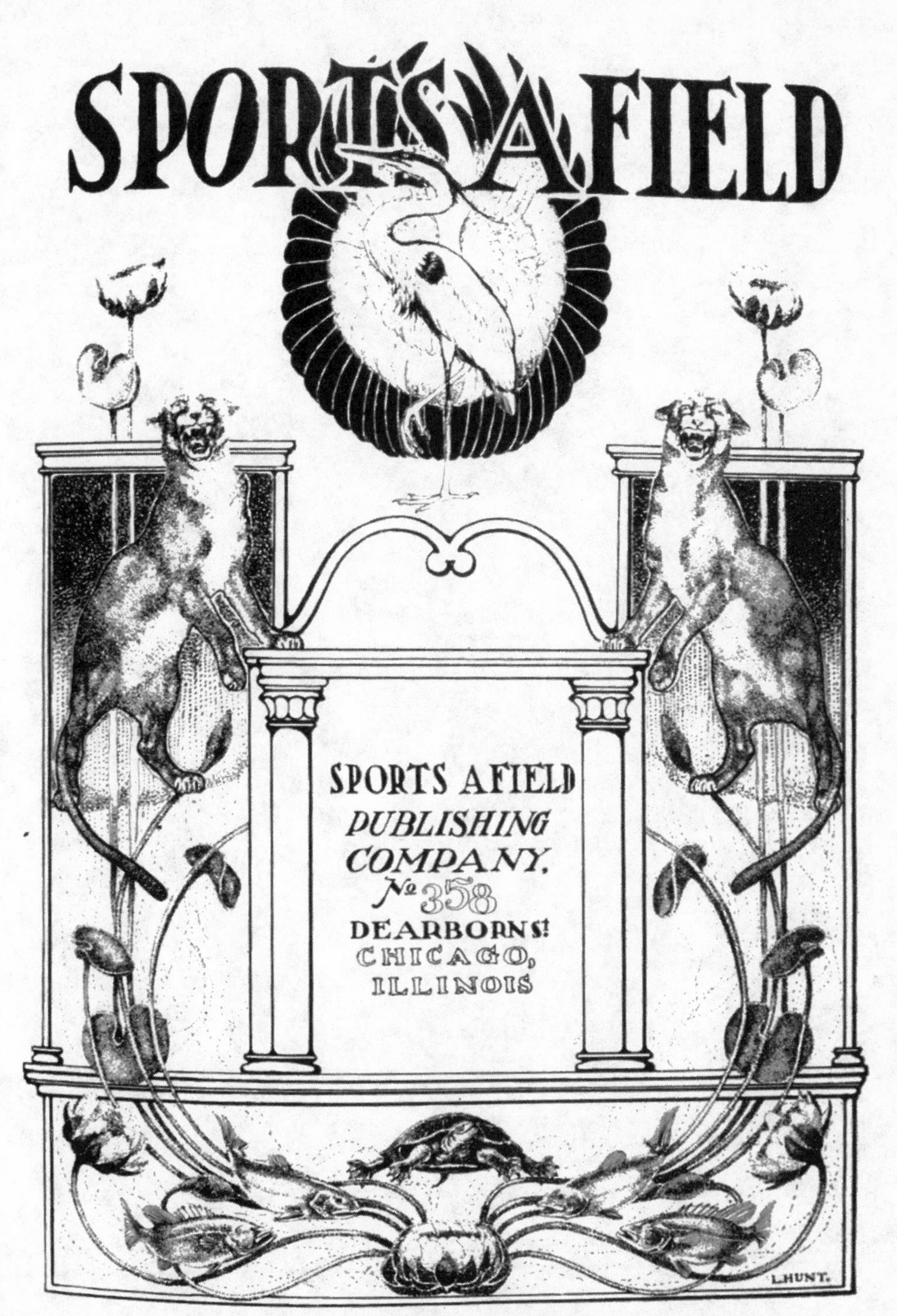

A New Serial. **THE RED ROBIN.** By S. D. Barnes.

IN APRIL ISSUE.

3
Guns, Ammo and Shooting

It is never advisable to buy a low-priced gun if you can afford a better one. You may not secure a commensurate return for the added expense, but the chances are that you will; and, in any case, there is to be considered the question of pride in one's possessions.

—J. F. Ehrens, *Sports Afield*, October 1901.

Deer-Gun Options

One consequence of the dramatic increase in deer numbers from coast to coast is that many hunters are systematically confronted with a decision: What gun shall I get for deer hunting? Many are youngsters just entering the age at which, legally or physically, they can hunt big game. Others may have hunted for years, but not for deer. Still others are nonhunting adults who have decided that they've been missing out on a good thing.

Depending upon where you hunt, some of the deer-gun decisions may have been made for you. In Pennsylvania, one of the top deer states, you can't use a semiautomatic rifle. In other states or portions of states, you can't use a rifle at all on a deer hunt.

Selecting a deer rifle can be especially confusing for those without exposure to centerfire calibers. The array of options is bewildering. Any gun catalog offers a variety of actions and calibers, bullet weights and barrel lengths, iron sights and scopes. Sorting it all out isn't easy without getting some competent advice.

1. Calibers: Deer calibers can be lumped into two practical categories: short range and long range. As short-range numbers, consider the 30-30, 30 Remington, 30-40 Krag, 35 Remington, 444 Marlin, 44 Rem Mag, 375 Winchester and 45-70. Almost all other calibers, including the good oldies such as the 250-3000, 300 Savage and 257 Roberts, are long-range calibers—those with consistent deer-taking capability beyond 150 yards.

Unless there is a compelling reason to buy one of the short-range calibers—an addiction to the 30-30 in a lever action is an excellent example—buy one from the long-range group. It will obviously anchor deer at eyeball distance even better than it will beyond 150 yards.

For virtually all beginning deer hunters, the choice of caliber should be on the conservative side. Select one of these five: 243 Winchester, 6mm Remington, 25-06 Winchester, 7mm-08 Remington or 30-30

Winchester (use the latter only in situations in which shots will be limited to 150 yards).

For the deer hunter who is a more experienced shooter, who can handle a bit more recoil, consider these four: 270 Win, 280 Remington, 308 Win and 30-06. The 30-06 deserves its reputation as one of the most effective all-around big-game calibers ever developed.

2. Shotguns: If you use rifled slugs, the most important consideration when selecting a shotgun is its sights. Decent accuracy isn't possible with "shotgun-type" beads. Fortunately, most manufacturers now offer slug barrels equipped with rifle sights, which is a giant step ahead in slug-shooting efficiency. As with rifles, a scope is even better.

Rifled-slug barrels that will usually give even better accuracy than factory smoothbore barrels are also available.

As for gauge, select the 12 gauge unless there are concrete reasons not to do so. In fact, nothing smaller than 20 gauge should be considered and that only for very modest ranges. The 10-gauge magnum, obviously, carries a much heavier payload of buckshot for those who can handle the weight and recoil.

3. Actions: Any action is "right" for deer hunting—in some places, in some hands. Good rifles are available in lever, bolt, semiauto, slide action and single shot; shotguns, in semiauto, slide action, single shot and doubles. A gas-operated autoloader is less punishing in recoil, making it particularly attractive to small shooters. A practical safety solution is to use it as a single shot until the novice is more experienced.

So don't buy a semiauto for a beginner, but don't hesitate to pick any of the other actions. And for those who are experienced shooters, the semiauto may be perfect.

A lightweight bolt action or a lever action makes an excellent beginner rifle. If you'd rather go with a shotgun, try a slide action or an autoloader used as a single shot. A double-barrel shotgun isn't a good choice for shooting rifled slugs, since few of them will shoot to the same point of aim with each barrel.

4. Sights: The open sights on most factory rifles aren't bad, and the bead sight on a shotgun is adequate for wingshooting. But they suffer terribly when compared to either aperture (peep) sights or scope sights for rifle or rifled slug accuracy.

Beginner Insights

The open sights on most factory rifles aren't bad, but they suffer terribly when compared to either aperture (peep) sights or scope sights. Aperture sights are very efficient if they are of hunting design (a big aperture, primarily); are properly installed (quite near the eye when in shooting position); and if the user practices. But for most deer hunters, a telescopic sight is best. Contrary to some beliefs, it is fast and easy to use. For the beginner, a fixed-power scope of no higher than 4X is perfect.

An aperture sight is very efficient if it is of hunting design (a big aperture, primarily); is properly installed (quite near the eye when in shooting position); and if the user practices sufficiently to become comfortable with it. But for most deer hunters, a telescopic sight is best. Contrary to some beliefs, it is fast and easy to use.

Barrel Length

Many gun experts say the minimum barrel length for high-powered rifles should be 22 inches. While it is true there is some loss in velocity for every inch of barrel loss, most hunting situations call for shots under 300 yards, and an 18 ½-inch barrel will certainly allow the projectile to retain more than enough velocity and energy to bring down the largest of big-game species. The key here, as in all hunting situations, is bullet placement and a sound knowledge of one's rifle and its capabilities.

Bullet Body Parts

✳ ***Base:*** The bullet's bottom.

✳ ***Cannelure:*** Cannelures are the flutings on Doric columns, and on bullets they are the grooves around their circumferences, used to identify them or to hold lubricant or for the case mouth to crimp into.

✳ ***Core:*** The lead, interior portion of a bullet that accounts for most of its weight. A metallic element called antimony is added to the lead as an alloy to increase the bullet's hardness and control its expansion. Manufacturers use various methods to bond or lock the core to the jacket, to prevent the two from separating on impact.

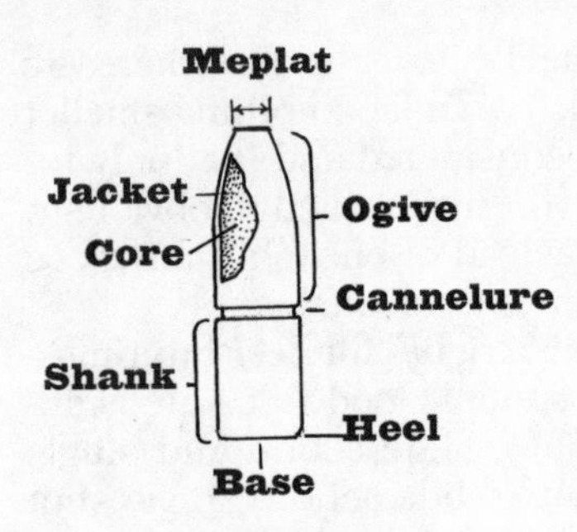

✳ ***Heel:*** The outer edge of a bullet's base.

✳ ***Jacket:*** The outer sheath of a bullet, made of gilding metal—usually copper alloyed with a small percentage of zinc. Bullet jackets arose with smokeless powders and the resulting high velocities: Solid lead bullets tended to foul the insides of the barrels, the lead deposits affecting accuracy. Jacket thickness helps control bullet expansion.

✳ ***Meplat:*** The diameter of the blunt end of the bullet's point. The size of the meplat is involved in figuring a bullet's form factor and thus its ballistic coefficient.

✳ ***Ogive:*** The curved, forward part of a bullet; the gradualness or sharpness of this curve, used for classifying form factor. "Ogive" is also the name of a pointed Gothic arch.

✳ ***Point:*** The tip of the bullet.

✳ ***Shank:*** The cylindrical bearing surface of the bullet beneath the ogive.

Bullet Types

✳ ***Boattail:*** Bullet with a pronounced tapered heel. This taper reduces aerodynamic drag, caused by the vacuum a bullet's passage through the air creates behind it, allowing a bullet to maintain its velocity over a long distance. Useful for long-range target shooting and military sniping, but of only marginal value at most sport hunting ranges.

✳ ***Flat Point:*** Similar to a round-nosed bullet, but with a flat tip. This is a good bullet for shorter ranges, and the flat point makes it safe for use in rifles with tubular magazines, preventing the possibility that the point of a bullet will detonate the primer of the bullet ahead of it in the magazine during recoil.

✳ ***Full Metal Jacket:*** The core is covered by a heavy jacket, usually of nickel silver or clad steel. A generally large-caliber bullet, for use only on the largest of dangerous African game.

✳ ***Hollow Point:*** A full-jacket bullet with a hole in the tip. Thin-jacketed, highly accurate and fast-expanding, hollow points make good varmint bullets.

✳ ***Round Nose:*** A bullet with a blunt, rounded point. Usually a heavy bullet, good for deep penetration at relatively close ranges.

✳ ***Spire Point:*** A bullet with a nearly straight-sided, conical, rather than a more sharply curved, front portion.

✳ ***Spitzer Point:*** A sharply pointed bullet—"spitzer" comes from the German word for point, *spitze*—specifically designed to be more aerodynamic and, therefore, more "ballistically efficient."

Bullet Shapes

Rifle bullets have come a long way in the last few decades, but a person can get buried in research trying to decide the best one for a particular hunting situation, in terms of ballistic coefficient, sectional density, etc. There are two basic rules you need to keep in mind. These pertain to how well a particular bullet will perform in the areas of retained energy, wind bucking and trajectory.

✳ ***Rule #1:*** Unless you are using a tubular-magazine rifle, always shoot a pointed bullet. This will out-

Trigger Jobs

If your rifle is completely stock, just like it came from the factory, it could probably use a trigger job. Whether you get a competent gunsmith to adjust the trigger pull to a lighter, cleaner release, or you replace the trigger with an after-market brand, you'll now be able to fire at the exact instant you want. You'll shoot more accurately, both on targets and on game, and you'll get more enjoyment from shooting.

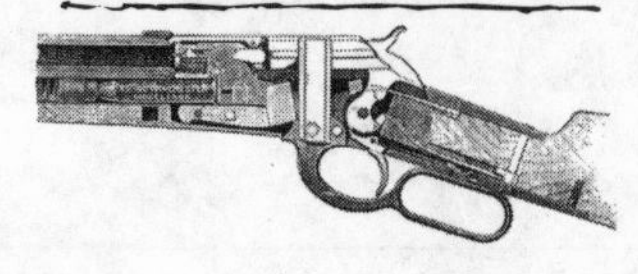

Ballistics for Selected Deer Calibers							
Weight		**Velocity (feet per second)**			**Energy (foot-pounds)**		
	Grains	**Muzzle**	**100 yds.**	**300 yds.**	**Muzzle**	**100 yds.**	**300 yds.**
243 Win	*100*	*2960*	*2697*	*2215*	*1945*	*1615*	*1089*
264 Win Mag	*140*	*3030*	*2782*	*2326*	*2854*	*2406*	*1682*
270 Win	*130*	*3060*	*2776*	*2259*	*2702*	*2225*	*1472*
7mm-08 Rem	*140*	*2860*	*2625*	*2189*	*2542*	*2142*	*1490*
280 Rem	*150*	*2970*	*2699*	*2203*	*2937*	*2426*	*1616*
7mm Rem Mag	*150*	*3110*	*2830*	*2320*	*3221*	*2667*	*1792*
30-30 Win	*170*	*2200*	*1895*	*1381*	*1827*	*1355*	*720*
308 Win	*150*	*2820*	*2533*	*2009*	*2648*	*2137*	*1344*
	180	*2620*	*2393*	*1974*	*2743*	*2288*	*1557*
30-06 Spr	*150*	*2910*	*2617*	*2083*	*2820*	*2281*	*1445*
	180	*2700*	*2348*	*1727*	*2913*	*2203*	*1192*
44 Rem Mag (20" barrel)	*240*	*1760*	*1380*	*970*	*1650*	*1015*	*501*

perform a round- or flat-nosed bullet, generally providing you with improved trajectory, more retained energy and less wind effect on its travel. There will be a difference from one bullet to another, but in general terms, pointed is best. However, in the case of tubular magazines, safety dictates a blunt-nosed bullet must be used. When the magazine is loaded, the point of the preceding bullet may contact the primer of the cartridge in front of it and accidentally discharge.

✳ ***Rule #2:*** Try to select a bullet weight that is common to the caliber. In other words, do not go to the extreme in weights. A bullet that is very light for the caliber, or very heavy, will not perform as well at long range.

How a Bullet Does Its Job

Kinetic energy (energy of motion) is the best measure of a bullet's killing power. However, proper bullet design is necessary to take full advantage of that power. The accompanying cross sections of a deer show four different types of bullet performance.

✳ ***A:*** An expanding bullet with a thin jacket and traveling at high velocity will disrupt with explosive

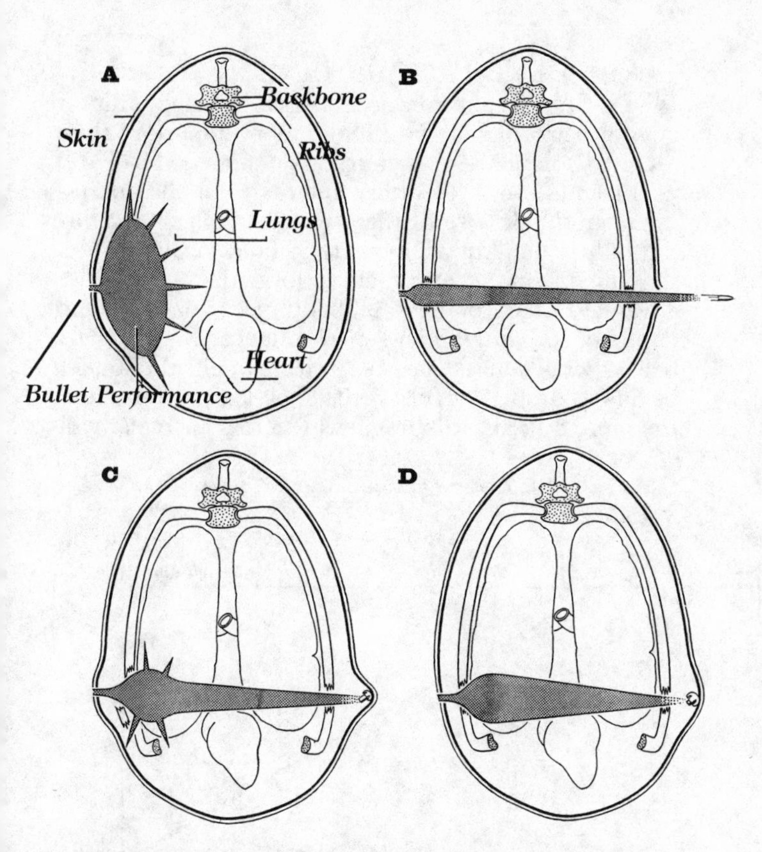

effect when it strikes. This would be fine for killing small varmints, but might badly wound a deer or other big-game animal.

✳ ***B:*** A bullet that is solid or otherwise too tough for the velocity at which it is traveling (such as a full-metal jacket) would not disrupt at all and would plow completely through an animal without doing substantial damage. The energy of the bullet as it emerges from the far side of the animal is wasted. It is possible that an expanding bullet would perform as in figure A at close range, when the velocity is high, but perform as in B at long range, when the velocity is low. This would, of course, be very unsatisfactory for killing big game.

✳ ***C and D:*** The ultimate in bullet action occurs when the bullet performs as in C at short range and D at long range. In both cases the expansion is controlled. The bullet gives up all of its working energy as it passes through the animal and comes to a rest under the hide on the far side.

Drop-Tables Explained

Of all the information on gun and ammunition performance, nothing is more important than drop figures. Of course velocities and ballistic coefficients are of technical interest, but the sportsman can't make use of them when it comes to sighting in his rifle, knowing where to hold, or how much to adjust for elevation at various distances. The advantage of a drop-table is that it gives you a graphic depiction of your bullet's performance. In the chart below, note the steepness with which the bullet begins to drop after 300 yards. The longer your shot, the more difficult it is to guess where your bullet will

A Deer-Caliber Drop-Table

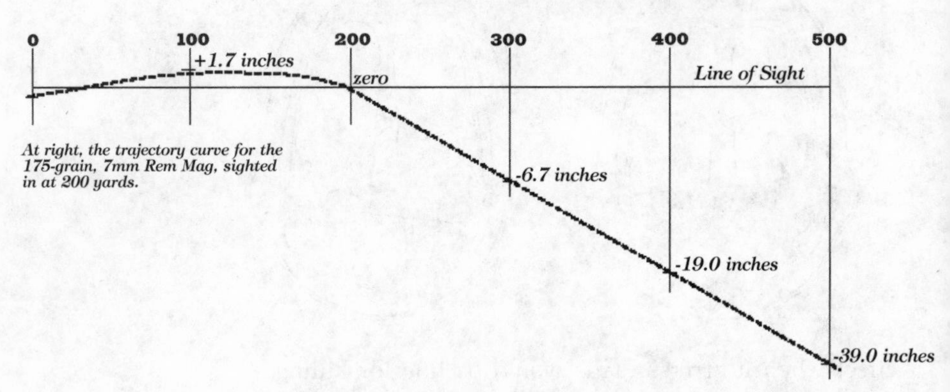

At right, the trajectory curve for the 175-grain, 7mm Rem Mag, sighted in at 200 yards.

hit. So—if you want a reliable sight setting for long-range shooting, be sure to sight your rifle in at the longer range, or at least obtain a drop-table from the manufacturer. Of course, most deer hunters will rarely, if ever, need to take a shot farther than 300 yards. If you expect to confine your shooting to short distances, sight your rifle in at 200 yards. You can then hold point blank out to 200 yards and never be off by more than about 1 ½ inches in sight setting.

Secrets of the 6.5mm

Not every rifleman has heard of the 6.5mm (264-caliber) cartridge. However, in Greece, Italy, Japan and Sweden, the 6.5mm is well known and highly respected for its accuracy. In the United States, the 6.5 x 55mm Swedish Mauser cartridge and the 6.5 x 50mm Japanese cartridge are probably the best known, and for good reason. In the 138- to 140-grain weights, the 6.5mm bullet has the best ballistic coefficient when compared with bullets of similar weights in 270, 7mm and 30 calibers. That means the 6.5mm retains both velocity and energy longer and also for a greater distance, all other things being equal.

If you are a handloading aficionado and are looking for an accurate cartridge with light recoil to use for hunting deer, this may be the one for you. The Swedish Model 38 and the Japanese Model 38 can each be purchased for around $200, and with a little talent and a minimal cash outlay, they can be converted into fine deer rifles.

Rainproof Rifles

Even the most tenacious of the silicone-based water repellents is apt to wash off over a day's hunt in a steady or heavy rain. A deer hunter I know favors car wax to protect his guns from the elements. He rubs it over the entire gun—stock and all—and then buffs it off with a soft cloth. Water simply beads up on the wax-protected surface. The wax job will outlast gun oils and spray lubricants. Just be careful to avoid waxes that contain buffing compounds, which could remove bluing or stock finish.

Best Bullet Weights for Deer

Be glad you don't design bullets for a living. A big-game bullet has to stand up to sudden, smashing pressures (when it is shoved down the barrel at 180,000 rpm), penetrate deeply and always expand (but never disintegrate). And it must be accurate.

Most bullets are accurate enough for deer. It's in the area of terminal ballistics—what happens when a bullet hits the target—that designers spend hundreds of hours.

At one end of the terminal ballistics spectrum is the varmint bullet, which is designed to fragment completely upon impact. Many varmints are small and offer little resistance, so the bullet's "expansion" must be rapid. Also, a bullet that fragments will not ricochet, which is an advantage. At the other end of the spectrum lies the "solid" bullet—really a full metal jacket—designed for deep penetration on large, dangerous game. Solids are not necessary (and generally not legal) for North American game, and certainly not for deer.

In the huge area between these two extremes we find controlled-expansion bullets. They start expanding upon impact, and are supposed to stop expanding before they disintegrate. A bullet that fails to perform properly usually expands too quickly or not at all. It's easy to avoid using the wrong bullet, though: If your rifle is 24 caliber or larger, the weight of the bullet will tell you if it's of a design suitable for deer.

In 24, 25 and 264 calibers, pick a bullet near the heaviest available. In 7mm through 30 caliber, pick one in the middle weights. In those calibers, the light bullets are made for varmints and the heaviest bullets are made for really large game. Few quick-expanding bullets are made larger than 30 caliber, so just about any bullet will do.

Recommended weights:

Caliber	Bullet Weight (in grains)
243	*95-100*
257	*100-120*
6.5mm (264)	*100-140*
270	*130-150*
7mm (284)	*140-160*
308	*150-180*
Above 30	*Any bullet weight*

The Surest Way to Learn Trajectory

Relatively few deer hunters know the bullet trajectory of their favorite deer cartridge. This is a primary cause for misses and poor hits on deer, particularly at longer ranges.

While close approximation of bullet trajectory can be learned from ammunition charts, each rifle has a somewhat different bullet trajectory curve. The best way to learn the true trajectory of your rifle and a particular load is first to make a well-centered five-shot group at the range at which your rifle is sighted in. Then, with the rifle well centered at this distance, using the same bull's-eye, shoot groups at 50-yard increments out to the maximum range you would expect to shoot.

When plotting trajectory, always shoot from a steady rest, and stay with the same brand of ammunition and bullet. Use a sufficiently large target to record all shots, and a safe bullet backstop. Another benefit is that you will learn the accuracy of your rifle at each given range.

How to Sight In Your Rifle

Before you can hunt or practice shooting successfully, you'll first need to sight in your rifle. If you've just mounted a new scope or purchased a new rifle, bore-sighting is the first order of business. Most gunsmiths have optical collimators that will do this job in seconds.

If you don't have access to a collimator, remove the

belt, cradle the rifle firmly in sandbags and look through the bore from the breech end. Move the rifle around until you can see a knot on a fence or some other easily visible reference point as you peer through the bore. Then adjust the scope's windage and elevation settings until the crosshairs intersect the same reference point. If you own a lever-action or autoloading rifle, you'll have to use a collimator.

The next step is to take the rifle to the range for some shooting. Be sure the rifle is solidly supported by sandbags, a rolled-up sleeping bag or some other steadying aid. Place the target 25 yards downrange, and fire at its center. If the shot is low, move the crosshairs up; if it strikes to the right of the bull's-eye, turn the windage-adjustment screw to the left.

Most rifle scopes are calibrated with half- or quarter-minute clicks; this means the bullet's point of impact moves a half inch or a quarter inch on a target 100 yards downrange with each click of the adjusting screw. When you fire from the 25-yard mark, you'll need to dial in four times the adjustment needed at 100 yards. To move the bullet strike each inch at 25 yards with quarter-minute increments, you'll need 4 x 4, or 16 clicks. At 100 yards, only four clicks are needed.

Once the bullet is striking near dead-center at 25 yards, move the target to the 100-yard mark. Fire three shots at this distance. The bullets should make a tight cloverleaf measuring no more than a couple of inches across. Once again, make the needed adjustments until the bullets strike where the crosshairs intersect.

If you're likely to encounter shots up to 250 yards, it's a good idea to zero the rifle so the bullets strike about 2 ½ inches above the bull's-eye when you center the target with the crosshairs. If you're using a modern, flat-shooting rifle, this will allow you to hold a deer-sized animal dead-on out to 250 yards—and hit the vital area.

Know Your Weapon Well

Most good shots at deer span only seconds. If you're prepared, that's time enough. Get to know your rifle by handling it often. Throughout the year, take time to lift it from the rack, open the action to ensure safety, then dry-fire at some spot on the wall. Do it again, "shooting" that spot twice, quickly. After a few hundred repetitions, your cheek will be sliding into the right spot on the stock with no effort, and your hand will naturally and unerringly find the bolt and the safety.

Airguns for Rifle Practice

Modern airguns have the heft and feel of a big-bore rifle and are extremely accurate at short range, making them an ideal choice for off-season practice for deer hunters who have no convenient place to shoot a centerfire rifle. A few hay bales make an adequate backstop for backyard shooting. There are even special traps to stop pellets, so an airgun can be fired inside a house or an apartment.

The top airguns are spring-piston operated. They cock with a single motion that pumps the air chamber and produces very consistent muzzle velocity, generally 600 to 1000 fps. A pneumatic pump system is less expensive but requires several pumps of the mechanism for maximum velocity.

If you use a scope on your deer rifle, equip the airgun with a similar scope, but be sure to buy one designed for air rifles. Oddly enough, scopes designed for centerfire rifles do not hold up well to the two-stage recoil developed by an airgun. Furthermore, centerfire scopes are designed to work best at 100 yards or so, whereas scopes made for airguns are constructed with 10-meter shooting in mind.

For $200 or less, you can buy an extremely accurate airgun-and-scope combination that can be shot for pennies a day (a box of 500 precision target pellets costs about $6).

Clean Cartridges

Many shooters are religious about cleaning their rifles but often neglect an equally important item—their ammo. Owners of semiautomatics often learn this lesson the hard way. Cartridges pick up all kinds of foreign matter, particularly if they are left in a leather cartridge holder for an extended period. When a dirty shell is rammed into the chamber by the bolt, it may stick just enough so that the extractor won't remove it, and a jam results. If you regularly clean your cartridges with a lightly oiled cloth, jamming will become a rarity. If not, you may not get a second shot at that big buck.

How to Hit Deer: 10 Tips

1. Sight it in. A few hunters still open the season carrying a new rifle that they've never fired, much less sighted in. Many veterans begin the season without making sure their rifles have maintained their zero over the previous 11 months. Whether you use a rifle or shotgun, be certain you know where it shoots.

2. Tighten the rifle screws. Make sure the screws on the firearm are tight—including those on the scope or the iron sights. They tend to loosen because of vibration, shocks and changes in humidity and temperature, often altering the point of impact.

3. Improve the trigger pull. Few procedures can improve shooting performance as easily as a good trigger job. Most factory rifles have pulls that are very heavy, and sometimes rough and creepy, but most trigger mechanisms can be adjusted by a competent gunsmith.

4. Know your trigger pull. You should know exactly when your firearm will fire. Ignore the old, erroneous advice of squeezing the trigger so gently that you don't know when the gun will go off.

5. Practice. Elementary advice, but the admonition to practice is a critical one. Nothing else prepares a hunter as does pulling the trigger, often on the firearm he'll use when the deer season arrives. Experiment with various shooting positions under hunting conditions. See how fast you can assume them, taking advantage of anything that could help steady your rifle for the shot.

6. Know where to shoot. Don't aim at the buck; aim at a very small portion of it. Decide exactly where your sights should be, from all possible shooting angles, to put the bullet into the vital heart-lung zone. Practice those sight pictures. Tape deer pictures on the wall, and use your deer rifle (unloaded, of course) to practice sighting on them.

7. Know when to aim low. When shooting at extreme angles either up or down, your rifle will shoot high. The reason? The horizontal distance, which determines the effect of gravity on bullet drop, is always less than the line-of-sight distance under those conditions. At medium to long range, that can be a significant factor.

8. Memorize the trajectory. Select the cartridge loading you will use for your deer hunting and then study its trajectory statistics. Memorize where the bullet will strike at key distances for your hunting

Never shoot at a running deer unless you are certain you can drop him.

conditions. Taping a simple chart to your stock is an excellent idea.

9. Remember wind drift. Wind drift is no problem for deer hunters where ranges are short, but for those where 200- to 400-yard opportunities are common, it can be disastrous. Most reloading manuals and some ammunition company trajectory tables provide drift figures at varying ranges for different wind velocities. Learn what the wind will do to your bullet, develop a feel for wind velocities, and adjust your point of aim accordingly when you make the shot.

10. Prepare for the shot: Play a mental game of "what if" while you're hunting. Continually visualize possible shooting opportunities, and evolve shooting scenarios to deal with them. Be ready to shoot. Develop a mind-set that will let you react quickly and properly. With all the elements in place before the deer appears, you'll seldom fail to make an effective shot.

Safe Handling

Carrying a rifle across your back on a sling leaves your hands free. When you're coming down a hill or mountain, though, carry a rifle across your chest. If you slip, you will fall with your back against the mountain and your feet pointed downhill to stop your slide. If you had your rifle across your back, gun and scope would take the brunt of the fall.

Tips for Improving Accuracy

Here are four factors all deer hunters should consider when trying to coax the best out of their guns.

1. Check the way the bolt action is fastened to its stock. Some models use two screws, others three. Drop the rifle gently into the stock, start all the screws, but don't tighten them down. Set the rifle on its butt and pull down on the barrel so that the rifle is as far back in the stock as it will go. Make sure the screws are not binding. Tighten the front screw enough to keep the rifle from slipping forward. Now put the rifle in a cradle or padded vise and, using a screwdriver that fits the slot, screw it as tight as you can. Next, tighten the rear screw; it should turn easily right up to the last half turn. If it doesn't, you may have a bedding problem putting unwanted tension on the barrel and action. The middle of a three-screw system needs only to be snug enough not to back out due to recoil.

Watching the Wind

No matter what type of firearms you are using, wind direction is one of the most crucial factors in hunting. An old Native American trick employed by today's bowhunters can be useful for rifle hunters as well. Tie a small feather to your firearm with a piece of light monofilament fishing line. Now you can easily tell which way the slightest breeze is blowing.

2. Rifles with moderately stiff barrels sometimes shoot better with those barrels floated—free of forearm pressure. Floating is easily accomplished by sanding out any high spots in the barrel channel, starting about two inches ahead of the recoil lug mortise. Light, thin barrels often give best results with a bit of upward pressure at the forearm tip. Installing a two-inch bedding pad in the front end of the barrel channel is a simple matter. It's a good idea to test with shims before making this change. The floated barrel is less affected by heat expansion, minor warping or the tight-sling shooting technique, which pulls the forearm away from the barrel.

3. When bullet holes walk up your target as the barrel heats up and expands, stock pressure is probably increasing. Often the path will be directly away from the pressure point. Temporarily floating the barrel by putting a shim behind the recoil lug will quickly determine whether this is your problem. If it is, you can either remove wood at the pressure point or float the barrel permanently. Should shots still walk with the barrel floated, chances are you have a problem such as unrelieved stress, and short of rebarreling, there isn't much you can do.

4. Buildups of jacket fouling and powder residue contribute to poor grouping. With particularly rough barrels (not unusual in new rifles), this can occur with as few as five or six shots. But as a general rule, accuracy will begin to decay after about 15 rounds in magnums or 20 rounds in smaller cases. New guns frequently get more accurate and foul less quickly after 100 rounds or so.

The Ups and Downs Of Wind and Temperature

Anyone who has shot a rifle knows what wind can do to a bullet's direction. According to the *Sierra Bullet Reloading Manual*, a 180-grain pointed bullet starting at 2700 fps from a 30-06 will blow six inches off course at 200 yards in a 20-mph crosswind. At 300 yards that 20-mph wind will take it 14 inches off the mark.

No matter how strong the wind is, bullet drift will depend on the direction the wind is coming from. If you're shooting right into the wind or it's at your back, it will have "no value" in relation to your bullet's trajectory. However, if it is blowing at right angles to your gun, it has full value, and a mere breeze could affect trajectory significantly.

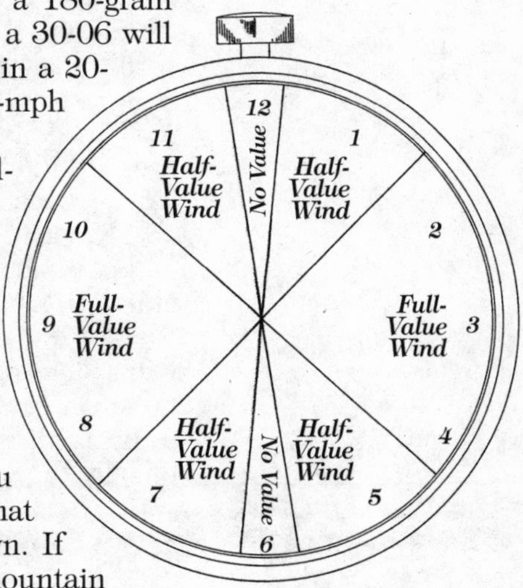

When hunting in the mountains, you need to pay attention to vertical winds that will cause the bullet to drift up or down. If you're shooting 300 yards from one mountain ridge to another through a 20-mph downdraft, a 30-06 bullet will drop several inches due to gravity and another 14 inches from the downdraft.

A 20°F temperature will make a 7mm bullet strike one inch lower at 200 yards and five inches lower at 500 yards if the rifle is sighted in for 100 yards.

The Partition Bullet

On contact, the forward portion of a partition bullet rolls back, thus greatly increasing its diameter. This increased diameter produces tissue cavitation and hydrostatic shock. The rear half of the partition bullet helps maintain bullet inertia, providing further penetration and killing power. Some manufacturers, such as Federal Cartridge, have found this bullet so reliable that they load it in their premium lines of ammunition.

Long-Range Shooting Ability

There is a considerable difference between theoretical ballistics and actual shooting performance, especially under field conditions. After practicing at a range, move into a field situation and test some of that ballistic theory. Pick a safe location and pace off 200 or 300 yards, whatever your ego tells you you can handle.

A non-returnable plastic gallon-sized milk or water jug makes an ideal target. It's about the size of a broadside heart-lung target zone. Fill a few jugs with water, adding several drops of food coloring in a color contrasting with that of the background. Then blaze away.

A direct hit will produce spectacular results, and even a crease will cause the fluid to drain away, producing a color change easily observable at extreme distances. If you don't connect after two shots, move closer in 20-yard intervals until you can hit consistently, thereby determining the true limits of your shooting ability.

While it may be a little humbling, it's far better to realize your limitations before the hunt than to be wondering about your ability after a long shot at dusk, with a faint blood trail leading into the woods.

Keep the First Shot Accurate

A bullet fired from a dirty bore sometimes impacts at a different spot than one fired from a clean bore. While many of us like to keep our guns as clean as possible, it may be better to carry a gun with a soiled bore in the field.

Very few of us clean our guns between each shot while sighting in on the range. Consequently we judge bullet impact against a history of shot groups fired from a "dirty" barrel. That first shot in hunting is usually the most important. If fired from a clean barrel, it may not impact where it did on the range. To solve this problem, you could run a

scrub brush down the bore between shots while practicing, but sometimes more than one shot is required in the field. A better answer is to fire a round through the bore just before going hunting.

It is important to know how much of an effect a clean bore will have on your point of impact in each of your firearms. Each will be a little different, and there will be times when a "fouling shot" may not be practical. Range-test each firearm when they are clean and soiled. By doing so you'll be better able to predict the consequences.

Do not use this as an excuse not to clean your guns. A dirty barrel will shorten the life of your firearms' accuracy, encourage rust and deterioration and, in extreme cases, be dangerous.

Firearms for Hidden Whitetails

Hunting for hidden whitetails is often an up-close experience. It is generally confined to small woodlots and thickets, and shots must be taken at short ranges. Therefore, hunters should choose guns and scopes that are most effective at 30- to 100-yard distances.

Any of the 30-caliber loads are adequate for this job, and they are more advantageous than the lighter, flatter-shooting cartridges. In this kind of cover, a hunter sometimes needs a slug that will steamroll through brush with a minimum of deflection. Good bullet-weight choices are 150 and 180 grain.

Scopes should be low in power and as bright as possible. Posts or heavy reticles are ideal because they offer better visibility in low-light situations.

The type of rifle action is a matter of preference. An autoloader or a similar type of fast repeater can be advantageous in the thickets. If a second shot is needed, it's usually needed in a hurry.

An extremely effective weapon for close-quarters deer is an accurate shotgun that shoots slugs. The weight of the slug, its energy and knockdown power make it a great choice for hidden whitetails. Don't, however, just take a gun off the rack, stuff in a couple of slugs, and go hunting with it. Accurate shotgunning for deer takes an open-choked barrel and several practice rounds to determine exactly where the gun shoots at various ranges.

The very best shotgun would be one with a special

Slip-Sliding Slings

Every hunter has experienced the problem of a slung rifle or shotgun slipping off his shoulder, but there is a simple solution. Sew a large button on the shoulder of your hunting jacket at the point where you want the strap to stop sliding outward. Buttons with convex sides are best because the edges are raised to catch the strap.

slug barrel and sights. A 1X or 2X scope adds tremendously to a shotgun's effectiveness, and it also provides for better low-light viewing.

How to Lose Your Concentration

✳ ***Phase One:*** During this initial stage of the hunt, the hunter maintains maximum concentration by thinking, looking, and listening only for deer. This phase usually occurs during the first hour or two, and the hunter is most likely to be successful in shooting a deer.

✳ ***Phase Two:*** The hunter's mind starts to wander from the topic at hand. As a result, the eyes and ears are less effective in detecting crucial sights and sounds. At this point, the hunter must stop and make a conscious effort to think about hunting and looking for deer. Stopping to "think deer" must be repeated continually throughout the day to maintain maximum concentration.

✳ ***Phase Three:*** As the mind wanders farther away from thinking about deer, looking and listening senses turn more and more toward the mind's rambling

Phase One

Listening | *Looking* | *Thinking*

Phase Two

Phase Three

Listening | *Looking* | *Thinking*

Phase Four

Listening | *Looking* | *Thinking*

thoughts, causing an increased breakdown in concentration. In other words, the hunter is daydreaming.

✳ ***Phase Four:*** The hunter is now merely going through the motions; there is a complete breakdown in concentration. Ears have shut out external sounds; sight turns inward. Any standing game is not observed. Unless the hunter can break this trancelike state, he should go home.

Scopes for Dual Drives

Although driving for deer doesn't require special equipment, certain options do offer distinct advantages, particularly in optics.

Since flushed deer are generally moving and often running, shooting is more difficult, and hunters have very little time in which to find an animal in the sights, aim, and shoot. A low-powered scope, or a variable scope with magnification turned down to about 2X or 3X, with crosshairs, is best for finding an aim point quickly. The hunter trying to pick out a moving target through a high-powered scope is bound to encounter one of two problems: Either the field of vision presented is so small that he cannot find the deer in the scope at all, or at close range, the hunter sees only brown hair in the scope and does not know which part of the deer he is looking at. A wide-angle scope offers a broader field of vision, which enables a hunter to spot a deer through the scope faster and follow a running animal easier, particularly when the power is turned down to about 3X.

As for rifle caliber, follow the cardinal rule of

How Telescope Optics Work

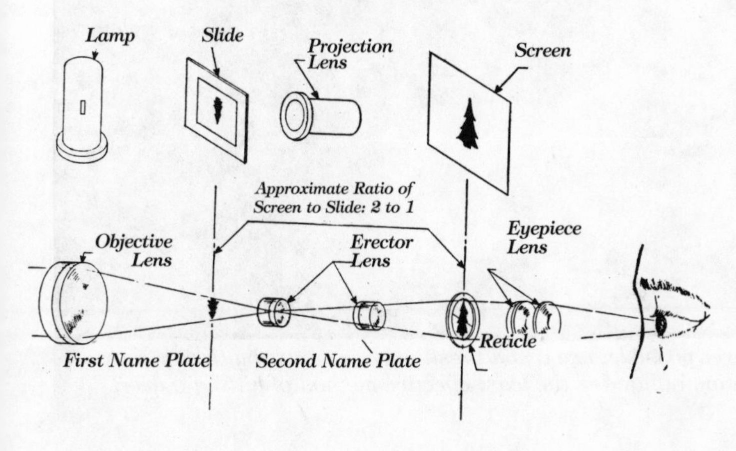

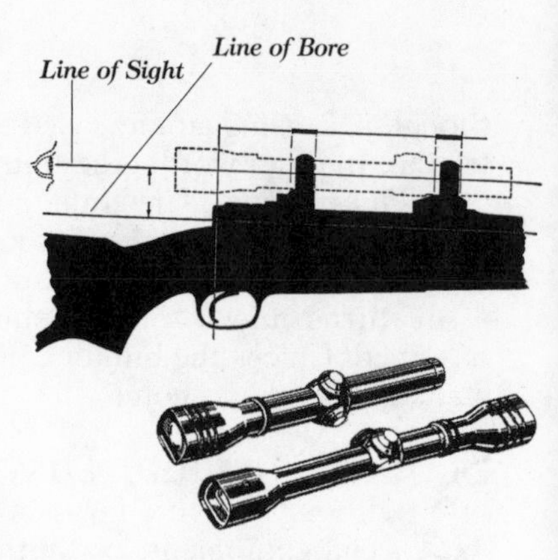

hunting: Shoot with what you have the most confidence in. However, using a light bullet (100 to 150 grains) that has a high velocity and flat trajectory will enhance accuracy and help eliminate guesswork. When shooting at a deer that is running across, it is necessary to lead the animal because it is still moving forward during the time it takes the bullet to reach it.

Five Things That Will Make You Miss

1. Failure to sight in before the hunt. Even if your rifle shot perfectly last season, you should sight it in again before you hunt this season and periodically during the season. Guns and scopes or even open

Whenever possible, use a solid rest, such as a nearby tree or rock. Shooting offhand is the least effective method of hitting a deer.

sights may get out of sync when they are hauled around on planes, trucks, horses or even when they are carried in the field. If you buy a new rifle, do not assume the bore-sighting technique used by the gunsmith who mounted the rifle scope has the scope perfectly aligned. Bore-sighting usually means a rifle will hit somewhere on the target at 100 yards.

2. Failure to use the same ammunition. Different bullet weights and even different brands of ammunition will shoot to a different point of impact in the same rifle. When you go on a hunt, pack plenty of your brand and bullet-weight ammunition, particularly if you shoot an unusual caliber or load that might be difficult to find.

3. Failure to use a solid rifle rest. Shooting offhand is the least efficient method of hitting your target. If there isn't a good rest in your vicinity, use your knee or, if possible, take the shot from the prone position.

4. Failure to practice. Most good rifle shots are made, not born.

5. Failure to recognize your limitations. Expert riflemen may routinely kill deer in the next zip code, but most of the 400-yard shots on deer are not nearly that far. If you don't feel confident you can make the shot, don't shoot.

Shortguns

Most hunters can shoot a shotgun better if both stock and barrel are shorter than "normal." Short stocks mount easily, and short barrels get on target quickly. Short barrels mean 26 inches for pumps and autos, 26 to 28 inches for doubles. A stock barely long enough to keep your shooting hand from recoiling into your face is about right.

Gun-Care Tips

1. A gun stored in a rack should be placed so the muzzle points downward at an acute angle. By pointing the muzzle down, excess oil is allowed to drain down the barrel instead of collecting dust in the action—not to mention the fact that it's a smart safety precaution.

2. Guns to be stored for long periods of time should be thoroughly packed with special grease made for this purpose. It prevents moisture from gathering on internal parts, which causes corrosion and pitting.

3. While there are many special gun oils on the market, a quart of regular 10-weight motor oil will provide more than adequate protection for firearms.

4. After a hunting trip where temperatures are below freezing, do not bring your gun into a warm cabin for the night. The metal parts will sweat when exposed to heat, and the moisture could freeze by morning, rendering the action useless.

Gun Transportation

How do you transport guns without damaging them and still keep them within easy reach? Here are some ideas:

*Plywood cases, made for gun transport, can be attached inside an RV closet or under a camper bed.

*Long steel drawers mounted in the cargo area of a sport utility vehicle.

*Window racks of pickups are popular with sportsmen, but they are not appropriate for long-distance travel. Fine hunting weapons will collect a lot of dust there and are always a tempting target for thieves along the way. Walk away from your vehicle for a moment, and one quick reach by a passerby can relieve you of a fine rifle.

*Van owners have the luxury of installing upright gun racks, covered with wood or metal sliding panels, in the tall rear cargo area. Incidentally, if the steel or wood rack boxes are removable, they can be lifted out intact and carried into a motel, cabin or RV at night.

Super Lube the Right Way

Most shooters misapply the modern "super-lubricants." They spray it on and immediately wipe it off. The lubricant, however, is most effective when given at least an overnight soaking period before being wiped off. This time allows the lubricant to enter the pores in the metal and drive moisture out. After a few of these treatments a firearm will shed water like a duck's back.

It's a Bore

Snow in a barrel can be a disaster. It stops the bullet or shot charge for a microsecond, pressure builds up, and the barrel bulges or bursts. Here are three ways to remove snow, or other obstructions, from a gun's barrel while you are in the field.

1. A Flexible Cleaning Rod: This useful accessory coils up for convenient storage. One end is threaded to receive a brush or patch rag. The flexible rod is the most reliable way to remove mud and snow from a rifle's barrel, and is very inexpensive.

2. The Sinker/Nail Trick: Attach a fishing sinker (for shotguns) or a heavy nail (for rifles) to a length of heavy monofilament line. When you drop the line down a barrel, it can be used to pull a bore brush through (Note: this technique does not work well if the barrel is plugged with mud.)

3. The Primer Trick: Cut a shotgun shell off above the brass head and remove all powder, wadding and shot. Insert the remaining casing into the chamber and fire. The slight pressure generated by the primer is usually enough to clear the bore but not bulge it.

U.M.C.

BIG GAME CARTRIDGES

THERE IS ALWAYS FRESH MEAT IN A U.M.C. CAMP

Your guide may know nothing of the U. M. C. factory with its automatic machinery, skilled mechanics and vigilant inspectors. He does know that U. M. C. cartridges do the work—are uniform, accurate and hard hitting. This oldest and largest cartridge factory makes every calibre from a B. B. for boys, to the largest brass cartridge case for cannon.

Used by the victorious American team.

The Union Metallic Cartridge Co.

Bridgeport, Conn.

Agency, 313 Broadway, New York City.

SAVAGE

HAMMERLESS TAKE DOWN RIFLE

¶ Simplest take-down, high power rifle on the market. Has all the strength, accuracy and endurance of the Savage regular '99 Model. ¶ Easy to take down as a shot gun; yet when assembled, the parts are automatically LOCKED into position. ¶ Can't be put together unless put together as tight and solid and rigid and accurate as a non-take-down rifle. Neither can it work loose by repeated taking down and assembling. ¶ Packs into small space; handy to clean, and loses none of its big game power by reason of its take-down feature. Examine this new Savage at all good dealers. ¶ Two lengths—22 and 26 inch, round barrels. ¶ Price, $20.00. ¶ Send for the new Savage catalogue. Every sportsman should have it.

SAVAGE ARMS CO.

6010 Savage Ave., Utica, N.Y.

Deer Rifle Overhaul

You want a new deer rifle, but you like your old one. No problem. Overhaul your old favorite by adding some new features. The costs range from $25 to several hundred dollars, depending on how much you want to do.

A hard-used deer gun needs proper care.

1. Stock Options: The most obvious place to start is the stock, and you have three ways to go. You can refinish the stock, get a new one of high-grade walnut, or restock with fiberglass or a similar material for a lighter, more durable rifle.

2. Rustproof Finishes: If you've subjected your gun to hard use, the bluing may have worn off in spots, or there may be some rust pits. A rebluing job will restore the luster your rifle had when it was new.

You can also make your rifle more weatherproof by treating the metal with a rust-preventive coating. Industrial-strength Teflon finishes not only rustproof your gun, but slick up the moving parts as well.

Electroless nickel is popular with many hunters. This silver-gray finish is rustproof and extremely tough. It does, however, give the action a "gritty" feel until it has been broken in.

3. Recoil Reducers: Experienced hunters accept recoil as a part of shooting centerfire rifles, but no one likes shooting the heavy kickers. All big-game rifles benefit from the recoil reduction a muzzle brake provides. With a brake, a 300 Magnum kicks like a 270, or even a 243. A 270 kicks hardly at all.

If you have a muzzle brake installed, be sure to get a thread protector you can screw on when you don't want to use the muzzle brake.

Don't forget to check the zero after you remove the muzzle brake. The point of impact usually moves in some direction, sometimes several inches at 100 yards.

On the end of the gun, opposite the muzzle, there's another opportunity to reduce recoil. If your gun doesn't have a recoil pad, put one on. If it does, consider replacing it with one of the new super-efficient models. These high-tech pads soak up recoil better than the traditional rubber pads.

Frozen Oil

Many owners of autoloading shotguns and rifles complain that their guns are useless in freezing weather because the actions freeze up. Admiral Richard E. Byrd, famed polar explorer, always cleaned his guns with benzene, and he made sure to remove all traces of oil from them, because even a small amount of oil will freeze an action solid in such extreme temperatures.

U.M.C.
BIG GAME CARTRIDGES

THE PALMA TROPHY

was recently won by the American Rifle Team, in competition with Great Britain, Canada and Australia. U. M. C. Cartridges were chosen to defend our National Honor after competitive tests with Government and other makes The winning world's record score was largely due to the superiority of U. M. C. Cartridges. Take them with you.

Look for the U. M. C. red ball—
the stamp of merit.

The Union Metallic Cartridge Co.
Bridgeport, Conn.
Agency, 313 Broadway, New York City.

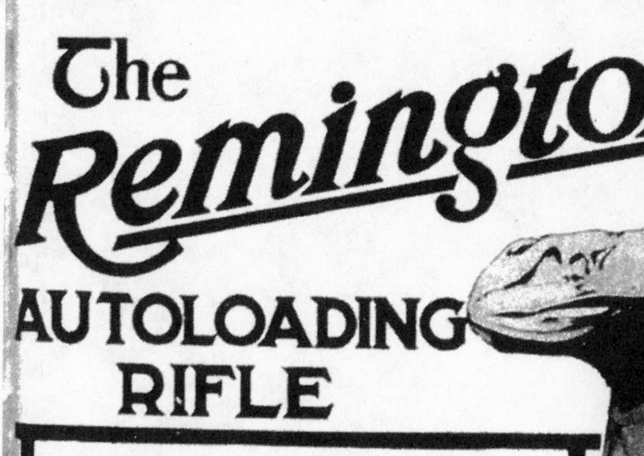

POWER—RAPIDITY—SAFETY

Five smashing, knock-down blows each equal to a ton hammer.
A lightning-like self-loading system—no lever and no slide to work.
A safe solid breech—a safe safety—a safe rifle.
.35, .32 and .30-30 Remington Autoloading calibres.
Compare it with other big game rifles, then

BUY A REMINGTON

Write Dept. 14 for New Illustrated Catalog.
REMINGTON ARMS COMPANY,
Ilion, N. Y.
Agency, 315 Broadway, New York City.

It loads itself and fires 5 SHOTS in one second

4
Hunting Tactics

Everybody in that country was a stranger to me, but an old gentleman by the name of Combs offered to board me if I could eat his kind of grub. When I found that he was another old hunter who could tell me just how to kill a deer I consented without a moment's hesitation.

—J. C. Banta, *Sports Afield*, September 1899.

How to Select the Right Stand Site

The single most important factor in stand hunting is where you place your stand. There are two phases to this crucial element of the hunt: 1) deciding on a general location to watch, and 2) choosing exactly where to place the stand in that area to get the best chance at a clear shot.

Here are some of the best options on stand locations. To pinpoint promising spots you'll need to spend time in the woods prior to the season, scouting for signs such as tracks, trails, beds, rubs, scrapes, droppings and the animals themselves. Then mark the possible sites on a map.

✳ ***Feeding Areas:*** Edges of cropfields, orchards, meadows, clearcuts or other areas with abundant forbs, weeds or mast. These are particularly good spots for bowhunters. Deer are less wary and less nocturnal in their feeding habits before major gun seasons open, and it's not uncommon to find bucks right out in fields of oats, corn or alfalfa during the first and last hours of daylight. The perimeters of major feed areas are also prime spots throughout gun season on lands with light hunting pressure.

Waterbucks

Since deer often travel along waterways, a top spot for a stand is on a high point overlooking the junction of two stream-bottoms.

✳ ***Trails:*** After feeding off and on during the night, deer usually head for thick, swampy or steep bedding cover shortly after daylight, then return to major feed areas in late afternoon. Locating stands along trails leading from feed areas to bedding spots is an excellent bet for a morning or afternoon hunt.

✳ ***Rubs:*** These are areas where bucks have scraped their antlers against the trunks of young trees. A single rub is not necessarily enough to warrant erecting a stand, but don't discount the importance of these marks, particularly if you find several fresh rubs in succession along a route that fits into a pattern of buck movement.

✳ ***Scrapes:*** These can be excellent locations just before or during the rut. Scrapes are oval areas where bucks have pawed away leaves and sticks and urinated over their tarsal glands, leaving their scent as a calling card to attract receptive does and to declare their presence to other bucks. Not every scrape will be visited regularly. Focus on fresh scrapes located in a cluster on a ridge or a field's edge.

Good spots for stands include 1) areas with numerous tracks, 2) edges of fields, 3) fresh rubs, 4) scrapes, 5) trails, 6) edges, 7) natural funnels, 8) spots downwind of heavily used areas, 9) creekbottoms, 10) areas littered with droppings.

✳ ***Natural Funnels:*** These are spots where the lay of the land forces deer to move through narrow passages. A funnel may be found where several ridges merge to form a single crest, where a strip of brush connects two fields, where rockslides or a river force deer along certain routes, or where two large patches of woods are joined by a ribbon of brushy cover in open terrain.

✳ ***Creekbottoms or Riverbottoms:*** Deer often travel

along stream edges. Walking is usually easy for them, cover and forbs are abundant, and crops are often grown because of the lush soils present.

✳ ***Edges:*** Always keep an eye out for areas where different types of vegetation join, such as brush and mature forest; or where different tree species, such as pine and hardwood, provide cover and several different kinds of food.

✳ ***Escape Trails, Cover:*** The opening of hunting season sends bucks fleeing for cover. Since most hunters are predictable in their approach and walk in along the easiest routes, you can use their movement, noise and scent to actually drive bucks to you. Study the lay of the land and determine rugged, thick or swampy areas where deer will go in search of a haven, then position yourself near the edge of that dense cover before first light on opening day.

The Long Stalk

When you spot a trophy animal in the distance, use binoculars to plan a stalk that takes advantage of all available cover and breaks in the terrain. Also be sure to avoid other animals between you and the trophy, so they won't spook him.

The Invisible Deer Stand

Whether hunting whitetails with a bow or gun, from either a tree stand or one at ground level, success in bagging a buck hinges largely upon your invisibility. Whitetails have tremendous visual capabilities and anything that is out of place usually catches their attention very quickly. They have one deficiency, however, in that they are colorblind, seeing everything in black and white, and this is something you can use to your advantage to ensure that both you and your stand are as inconspicuous as possible.

After you've decided on a stand location, and either built a semi-permanent blind or installed a portable stand, go fetch a buddy and ask him to bring along his instant-developing camera loaded with black-and-white film. A regular camera will also work if you have sufficient time to get the film developed. Now, climb into your tree stand with your gun or bow, dressed the way you will appear during the season.

Have your pal snap pictures of you in your stand, from various angles and directions. These pictures will give you a good representation of how you and your stand will appear to deer. If your entire presentation is easily recognizable, change the position of your stand to take advantage of some nearby heavy-cover backdrop or add a leafy branch to break up your silhouette.

You should be especially concerned with the many

light and dark tonal values exhibited by your stand since it's imperative that they match the tonal values of the cover surrounding you. Therefore, if the photos reveal overly light or overly dark renditions, or perhaps glare or reflection, you should tend to them before the deer season opens. One way is to use drab-colored spray paints. Or you can smear mud on your stand. Or you can thumbtack in place a yard or two of camo cloth whenever it is needed (remember to remove the tacks at the end of the season). As to your person and your clothing, strive to eliminate shine and wear several different tonal values. In those states that require fluorescent orange clothing, consider using the camouflage of mottled blaze-orange design.

Where Mountain Deer Hide

In mountain country, don't look for too many bucks on the top of ridge spines, because deer tend to travel and bed a short ways down from the top, where they are better protected and do not stand out against the sky.

How to Scan On Stand

The best way to scan for deer when you're on stand is to rotate your head slowly, studying the woods for small patches of color, shape or motion that could spell whitetail. Usually you can cover 270 degrees or more just by swiveling your head. Be alert for sounds; often, you'll hear a snapping branch or the rhythmic footsteps of a deer before you see it.

When you spot your quarry—if it is in range, presents a good shot angle, and has not yet seen you—smoothly raise your bow or gun and fire. If the deer is looking in your direction, freeze until it goes back to feeding or walking or turns its head away, then make your move. Bowhunters should wait 15 minutes to an hour before going after a hit deer. With a firearm, a few minutes' wait is sufficient if the shot was well-placed.

Raise your rifle slowly when you spot a deer.

When Not to Take a Stand

As effective as stand hunting is, there are times when other methods of hunting may be a better choice. Here are conditions that may dictate another approach.

1. Heavy hunting pressure has driven deer into areas of thick cover.

2. Deer have become nocturnal and have changed their feeding habits.

3. Unseasonably warm weather has slowed deer movement to a minimum.

4. Strong winds have stopped deer activity.

5. Hard rain or heavy snow has deer bedded up.

6. Pressure from feral or free-ranging dogs has pushed deer into swamps and thick areas.

7. Pressure from nonhunting human activity such as timber cutting or mining has pushed deer into thick areas or caused them to move at night.

8. Weather is just too cold to sit for hours.

Easy on the Scent

If you use prepared scent, you don't have to pour it all over your jacket for it to be effective. Instead, soak a piece of cloth with the scent lure and pin it to your shoulder pad. When you've finished hunting, you can eliminate the odor simply by taking off the flap.

Sit Tight on Opening Day

For the lone hunter, still-hunting on opening day, when the woods are generally in a state of extreme flux, is a rather dubious exercise. Far better for him to find a place in the preseason where deer normally move past in the mornings, and to set up there with a ground or tree stand. He can then hunt the deer's regular movements (such as from feed to resting cover) before those movements become anything but regular. Such a strategy would obviously call for a hunter's being on his stand well before daylight, and would probably result in his taking a deer early in the morning before matters get too hectic around him. But even if it's getting on toward midmorning and he's seen nary a deer, a hunter should nonetheless resist that urge to abandon his stand and see what he can sneak up on. On opening morning, if you know you've got a good stand, stay on it as long as possible—then stay another 30 minutes.

Post-Rut Scents

One tactic that works amazingly well during the post-rut is the use of sexual scents to attract bucks. In fact, estrus scents seem to work better now than they did during the primary rut, especially when applied from a drip dispenser attached to your boot. By wearing a dispenser, you can walk around your hunting location and create mock estrus trails of does. Such trails can attract bucks that might not have come near otherwise. Another post-rut strategy is to imitate the blat of a doe in heat. This deer call often brings bucks in on the run.

A Stand-Hunter's Checklist

1. Carry a Thermos of hot soup, tea or coffee.

2. Bring pruning shears to trim away branches that obstruct your shooting lanes. (Leave nearby branches that aren't in the way, to help camouflage your stand.)

3. Bring a rangefinder to determine the distance to

certain fixed objects. You can then judge how far the deer is by its proximity to these objects.

4. Bring a cushion for your seat if it's not padded.

5. Wear a masking scent such as fox or skunk, or put some scent in cotton in an open film canister and place it near the base of your stand.

Practice Your Stand Shot

Shooting from a tree stand is very different from shooting at ground level. You'll be more accurate and feel more comfortable if you practice from an elevated position before the hunting season arrives. Set your stand up in a safe area at heights you expect to shoot from, and practice shots at various ranges that might be typical for the habitat you hunt. For bowhunters, this might mean shooting from as far off as 35 yards to almost directly below the stand. A gun hunter might want to practice shooting out as far as 100 yards from his stand, as well as at closer ranges.

For gun hunting, virtually no compensation for angle is required in your aim point, because the angle from a 15- or 20-foot-high stand out to the deer isn't particularly sharp. Simply aim right at the vital area.

A bowhunter, however, usually must shoot lower than would be expected, to compensate for the sharp downward angle. New sights built especially for bowhunting from tree stands are available that compensate for this angle automatically.

Hidden Bucks

Don't expect to get a full view of a deer when still-hunting. Instead, look for parts of the animal showing between brushpiles, or sticking out from behind a tree trunk. Search for the glint of sunlight on a buck's antler, the white hair patch of a deer's throat, a black nose, a flickering ear or tail.

Drive, He Said

Many whitetail hunters tend to view drives as desperation measures better left for the end of the season. Frankly, if deer drives are going to be used, then it might as well be on opening morning. If your preseason scouting has shown you which patches of woods are holding bucks, then opening morning, before other hunters have had a chance to chase them around and disrupt the deer's patterns, is probably one of the best times to stage a drive. Studies have shown that whitetails' basic fall pattern is to bed in thick cover during daylight—often at the end of their home range opposite from their feeding area and, in hill country, on high ground rather than low—so the midmorning to midafternoon hours of opening day would seem the most logical period for driving deer.

When looking for signs of bucks in the preseason prior to the peak of the rut, look for fresh rubs and beds. New snow is a tremendous aid in locating deer's daylight bedding areas. Traveling farm or logging roads before sunrise and after a night's snow can show you exactly where the deer are headed after their nocturnal feeding.

Stalk the Seams

Edges are great locations for deer stands. The border of mature forest and a clear-cut or power line is a good choice. So is the edge of a pine forest and deciduous trees, or the edge of a cultivated field and a brushy area.

Hunting Around the Rut

✻ ***The Pre-Rut:*** Where legal and safe, take the tail of a deer and sprinkle it with buck urine and estrus urine. (This works especially well during bow season.) Thread 30 yards of string (any color but white) through a hole at the base of the tail. Hang the tail from a branch about 20 yards from your stand, at the height of a doe's rump (24 to 32 inches). Unravel the string from the tail back to your stand. Tie the remaining string end around your foot (or forearm, if you're in a ground stand). Twitch the tail every so often, while making a soft, long blat on a deer call.

This will attract bucks via their senses of smell, sight and hearing. Bucks respond because the rut has

A peak-of-the-rut buck urinating on its tarsal glands.

just started and they have few apprehensions. Just the twitching of the scent-laden tail is often enough to make a buck charge out of cover to investigate what he believes to be a hot doe. Calling along with this can attract bucks that are out of visual range, but have picked up the pheromones emanating from the tail.

✱ ***The Peak of the Rut:*** Unorthodox but effective, "rub-grunting" combines the pheromones of agitative scents that are common at this time of the year (tarsal and buck urine) with grunting. The pheromones stimulate bucks into responding. During the day, bucks generally wind-check their scrapes from the protection of cover. If a buck picks up an aroma he likes, such as that from a doe, he usually waits until she leaves the scrape and enters cover before he pursues her. If he smells what he thinks is a competitive buck, however, he often charges from cover to confront him.

To rub-grunt, locate a fresh primary scrape (two to three feet in diameter, in moist, bare earth). Set up 10 to 30 yards from it. Don't worry about making too much noise. I find that creating natural noises (snapping twigs, making soft grunts, shaking surrounding brush) on my way to the stand helps to set up the illusion of a buck invading another buck's primary scrape. Begin by actually pawing the leaves away from your stand, methodically and purposefully, as if you were making a scrape.

Next, deposit a combination of tarsal and buck urine scents onto the bare earth. Do not use too much, as a buck may be bedded down fairly close by. Overusing scent could alarm rather than attract him. While depositing the scent, make three to five burp-like grunts. These grunts must be very short, to simulate the sounds of a frustrated buck—*erp . . . erp . . . erp*. Blow the grunt sequence every three to five minutes, starting from when you paw out the leaves and disperse the scent.

Now comes the coup de grace. Take a small "Y" or forked-horn antler and vigorously rub a nearby sapling. Remember, to call a deer effectively, you want to create not only the entire illusion, but a realistic one as well. Bucks that rub saplings do so in a deliberate way. They rub their antlers several times, stop, sniff and lick the sapling or branch, then begin rubbing again. So take a break every eight or 10 rubs, as opposed to constantly rubbing. After rubbing the tree bare of bark, hesitate for a few minutes, then make a

Silent and Dry

Avoid wearing noisy foul-weather gear when still-hunting or stalking game, even if it's raining. Wear a wool shirt over the rain jacket, if necessary, to cut down on the sounds that can spook game.

Hunt Travel Zones

If you've located a good buck before the season, resist the urge to hunt him in his bedding site. This is almost certain to drive a wary trophy animal out of the area. Instead, hunt the travel zones between his bedding and feeding spots.

series of grunts. Shake the sapling aggressively, then start all over again.

Bucks are attracted not only by the competitive scent of the tarsal and buck urine, but also by the sounds and sights you have created with the scraped leaves, rubbed antlers and shaking brush. Be ready for bucks to appear any time when doing this.

✱ ***The Post-Rut:*** One post-rut method uses a commercially manufactured scent that combines doe-in-estrus scent mixed with the urine of an immature buck (2 ½ years old or younger), and it is deadly in areas that does frequent. Apply the scent to a boot pad and walk along the fringes of the main deer trail, heading toward your stand. Once you are 10 to 30 yards from your stand, walk a circle around your location and remove the pad. Sprinkle the sole of your boot with a couple of drops of estrus scent and go directly to your stand (this will help cover the human odor on your boots). To complete the strategy, make several long, whining doe bleats. This imitates what occurs in the woods: Does that have not been bred announce their "readiness" by walking through the woods while emitting whiny bleats, trying to attract bucks.

By making a mock estrus trail, you create first-come, first-served competition. A buck traveling down your mock scent trail will smell not only the doe's estrus, but the aroma of an immature, competitive buck as well. His reaction will be to trot along the trail looking for the doe, thinking it will be easy to run off the younger buck. Your estrus bleats serve to further stimulate the buck. Be prepared, as most bucks will be moving quickly. They may trot past your stand several times, trying not only to zero in on the doe, but to locate the other buck as well.

Southern Bucks

After bitter winter weather arrives, look for deer on south-facing slopes, where they can escape harsh north winds and soak in the warmth of the sun.

Mulies vs. Whitetails: To Stand or Not to Stand

One of the fundamental differences between hunting mule deer and hunting whitetails during the rut is in the method: mulies are generally still-hunted; whitetails are stand-hunted. For the most part this is due to the difference in the terrains the deer occupy. By the time of the rut, in the northern reaches, the mule deer will probably have migrated down into their winter range, chosen for its better feed and weather conditions. The whitetails don't move much at all from their summer homes.

For mule deer, then, in mountainous country, winter range is often large, sheltered bowls or sidehills, the mule deer showing an affinity for somewhat open spaces from which they can sight the approach of any threat. Whitetails, on the other hand, pass most of the year in tangled bottomland or wooded hills where they can steal along. So whitetail hunting almost demands that a hunter take some sort of stand.

If there could be said to be any shorthand formula for hunting mule deer in the rut, it might be to know what country they will be occupying and go looking for them in it; for whitetails, know their habits and let them come looking for you. (That is just precisely as short and foolproof a formula as it sounds. Hunting is such a woefully unreliable science that it is not a science at all. It is in all ways an art, a skilled art).

Deer in the Rain

Pine and cedar thickets are great places to look for deer during heavy rain, snow and wind. Deer can stay almost dry in these locations and escape the brunt of strong breezes.

Tips for a Creekbed Deer Hunt

1. Your outer clothing should blend as well as possible into the background, and this means a mixture

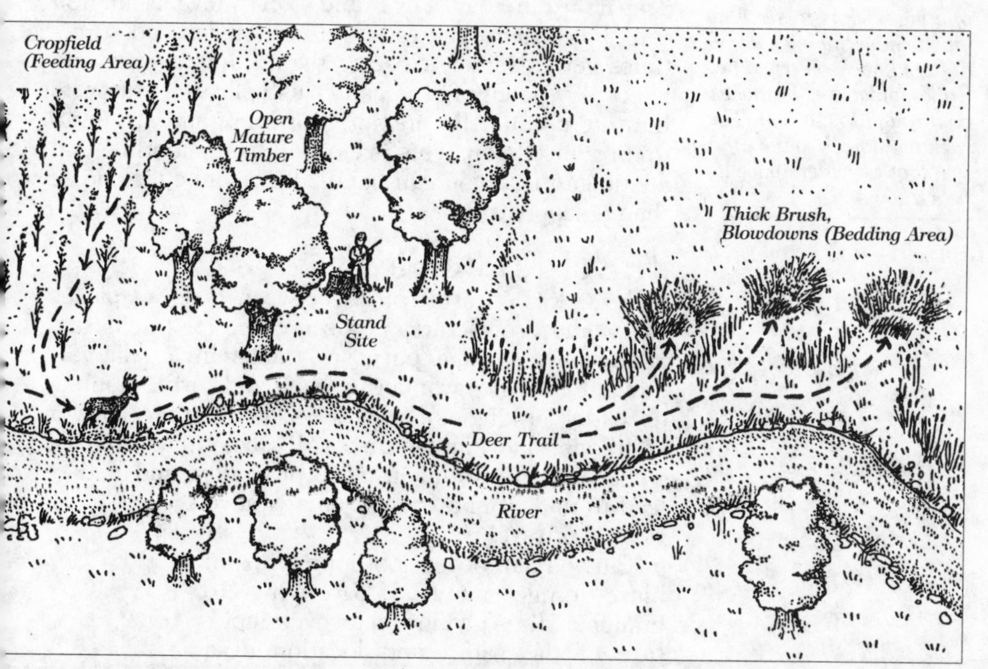

A good spot for a stand in the early morning and at twilight is near a deer trail along the edge of a riverbottom. Try to set up where you can intercept the deer as they move between their bedding cover and the nearest feeding area.

of tones, such as a muted plaid jacket or camouflage shirt. (Although deer are color-blind, they can see the light reflected from bright colors, as well as blocks of solid dark colors).

2. It's doubtful the firearm you use will require telescopic sights when hunting creekbeds because virtually all shots will be at very close range, where optics are a hindrance instead of an aid. Open-iron sights are your best bet.

3. You may see a lone trophy buck, but more often he'll be in a herd of three or more deer, so be aware that movement required to raise and aim may be seen by more than one pair of eyes, and choose your moment accordingly.

Windy-Day Bucks

Although many deer hunters don't like to go out on windy days, such days can often be good for stalking. When the wind is blowing hard, deer lie up in heavy cover and won't move at all. You can stalk brushy draws and other tangles and sometimes get extremely close to them. Move slowly and kick every patch of brush and slash pile as if you were hunting rabbits. Be prepared for a big buck to jump up in front of you. This is close-in, fast shooting, and a slug gun that you can handle fast is a better choice than a scoped rifle.

Monsters in the Snow

One way to tell if you're following the track of a big buck is to see how much snow the buck's antlers knock off when he walks under tree branches or through low brush. The rack will knock snow off a two-foot or wider space if he's a big one.

Travel-Route Savvy

Look for travel routes between bedding places and browse areas by searching for tracks laid down with a purpose—tracks in a relatively straight line, not wandering back and forth. Double-check by looking as far as 25 yards to either side for additional and similar tracks. If tracks made by the same deer are found in close proximity, and if they lead to and from a browse area, you've located a travel route.

Carry a half-dozen foot-long strips of blaze-orange material for tying to branches, thus giving you an overall picture of the width and location of a travel route. Once the route is firmly established in your mind, remove the strips—they could frighten deer.

Try to be on a travel route early in the morning, preferably about 10 or 15 minutes before good light. Wait quietly. Once it is shooting time, you are already in position for bucks that are on their way from browse to bedding. Set up near the buck's destination, not his starting point. For the afternoon hunt, reverse your stand so that again you are near his destination.

If you are still-hunting, start at the buck's destination and work toward the oncoming animals. Never hunt down the middle of a travel route. Stay to one side—whichever side is best from a cover and wind standpoint.

I always take two or three steps, then glass the area ahead. An average shot is 75 yards.

How to Hunt Islands

Some of the most productive deer hunting around water can be found on islands out in lakes or rivers. Depending on the size of the island, deer may actually live on it, or simply take refuge there when pressured by a lot of hunters.

One advantage of island hunting, of course, is that the deer have only a limited area in which to hide. Don't be fooled by this, however, since whitetails have an uncanny knack for becoming invisible.

If you are by yourself, slow, careful stalk-hunting is probably the best tactic to use. Try to crisscross the island, or let the natural features of the terrain guide you. If you are with a party of hunters, a small drive may be the best tactic, starting at one end of the island and sweeping across.

During these types of drives, you might walk right beside deer hiding in thick cover. Make sure

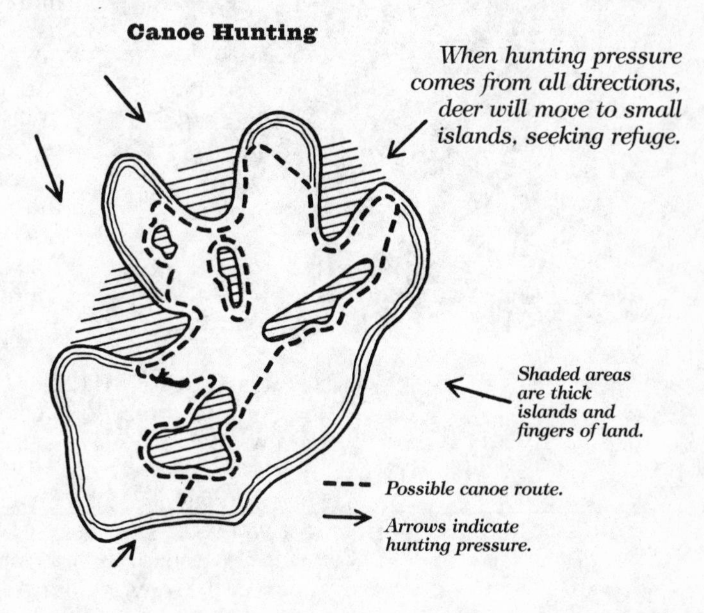
Canoe Hunting
When hunting pressure comes from all directions, deer will move to small islands, seeking refuge.
Shaded areas are thick islands and fingers of land.
Possible canoe route.
Arrows indicate hunting pressure.

you investigate such places thoroughly, and be prepared for a quick shot.

If the island is large enough, you may elect to sit and watch, either from a tree stand or a ground blind. Look for a well-used trail, an active scrape or perhaps a feeding or bedding area, and set up accordingly.

Make certain you know where you're hunting. Many river islands are private property, while others have been designated state or national wildlife sanctuaries.

Local wildlife officials and game wardens can provide specific information on the legality of hunting river and lake islands, so be sure to contact them before going afield.

From the Mouths of Bucks

Unlike many kinds of mouth-blown devices designed to lure wild animals, grunt calls for whitetails are easy to master. The dominant grunt is a very brief, low-pitched "ugh" sound that, with many of the commercial calls, can be imitated by saying the word "who" as you blow the call.

The coaxing or "tending" grunt used by a buck chasing an estrous doe is a more drawn-out sound, usually about a second in duration. It comes out sounding something like "Whoahah," the upward inflection on the end made by clamping a hand over the end of the call and then releasing tension as the coaxing, persistent grunt is sounded.

Walk Like a Whitetail

It is not, of course, so much a matter of going gentle into any good night as it is going quiet through the dry leaves of early whitetail season. Walking takes on unimagined complications when it comes to the deer woods.

During those periods of greatest deer activity early and late in the day, walking is not even the best course of action for whitetail hunters; climbing up a tree and squatting there in silence is. A basic theorem of deer hunting is that when deer are on the move, a hunter should sit still; when deer stop moving around, a hunter should move. So when the deer bed down at midmorning, that's when you should climb down from your tree stand. But too often what greets you on the ground is a blanket of noisy dry leaves.

One thing that can be to your advantage is a day of high winds. At such times the deer will generally hold tight, though they will also be exceptionally nervous and alert because of all the wind noise around them. By moving into the wind, keeping the sun at your back if possible, you stand a good chance of seeing a deer before it sees you.

What about days when the wind isn't blowing at all? You may still want to climb down and have a look around, even though walking silently is a near impossibility. The closest thing to a solution is to try to walk like a whitetail.

The human gait is a lively, rhythmic shuffle that says only one thing to deer who hear it: Yikes! Aside from the sound problem, such a walk does not give a hunter enough time to look over everything with due care. The normal, undisturbed gait of a deer, however, is an erratic cadence of distinct steps: three steps, stop, listen, two steps, stop, listen, four steps, stop, listen. The sound of this sort of walking is one that deer hear all the time, and one that does not immediately draw their attention.

Sneaking Through Crunchy Snow

The trick to fairly silent stalking through crusted snow is to follow someone or something else that has sometime previously broken a trail. Step in their tracks. Where they walked, the snow is compacted, minimizing the noise you are going to make as you hunt through. In areas where hunting pressure is medium to heavy, any number of good paths should be available to you. Where pressure is light, look for routes taken by a herd of deer, or elk, and follow them. With the snow already broken up, some degree of quietness is possible.

Quiet Boots

Quiet boots make quiet walkers, and rubber-bottomed pacs are probably the quietest of all. These boots give a hunter a "feel" of the forest floor much like that he would get by wearing a pair of moccasins, while the rubber bottoms protect his feet and prevent him from laying down scent. The original of such pacs is the Maine Hunting Shoe, first produced in 1911 when Leon Leonwood Bean, dissatisfied with all-leather boots for hunting, had a cobbler stitch 7 ½-inch leather uppers to a pair of ordinary rubber overshoes. Of the first 100 pairs Bean sold, 90 were returned because the rubber was not strong enough. Improvements—such as the full chain-tread sole—were made, and the rest, as they say, is history.

Floating for Venison

In the 1880s, floating was an extremely popular method of nighttime hunting throughout the Adirondacks. A lone hunter could be quite successful using this method; however, it required a high degree of skill at handling the canoe or guide boat and considerable shooting expertise.

The hunter would quietly paddle the waters of a marsh, lake or river until he heard the sounds of a wading deer feeding on water plants, a favorite food. He would then quietly direct his craft toward the sound and float or drift silently until the animal's outline could be seen in the darkness.

The technique saved pioneers many hours of trekking into the dense wilderness. It also allowed hard-pressed families to hunt after sunset, when the chores were done.

A Buck Scrape in Your Pocket

An artificial buck scrape can be a surefire way to bag one, and all the ingredients to make it can be carried in your pocket.

A simple way to make an artificial scrape is to bury a 35mm film canister containing doe-in-heat lure right at ground level. Fill the container with a piece of rolled-up cotton cloth or cotton balls to serve as a wick for liquid scent. Place a perforated cap over the container to keep dirt out and control the evaporation process.

To carry the scent bomb afield, neat and clean, just place the hermetically sealed lid over the container. Place your scrape under an overhanging branch. Carry handy drip bottles of doe urine in your pocket to refill and freshen the canister during hunting as you frequent the scrape. Scent bombs should last about five days between freshening.

I have had deer freshen these scrapes regularly after less than a week in use, and many will repeat from one year to the next.

Rattling the Rut

Rattling works best for drawing in whitetails just before or after the peak of the rut. Grunting on a tube call, however, can be effective year-round because bucks and does always use this sound for communication.

Tactics for Cold-Weather Mulies

Fresh snow is particularly valuable for late season hunting, since it allows you to see where deer have been moving. Search as far ahead as visibility allows while you trail the animal; you'll want to spot it across a draw or coulee before you get so close that it knows you're following it. Also watch to the sides of the trail you're on, particularly the uphill side, since the deer may sense you're behind and circle back to get a look at you.

Track fast if the buck is moving at a steady pace. If the deer's tracks start to amble somewhat from side to side and the tracks grow closer together, the animal is likely slackening its pace and ready to search for an area to bed down. Slow down and look ahead beneath trees and next to rocks for the prone gray form of the buck.

Try not to track a mule deer too closely; he will spot you.

The First Deer Hunters

Indian techniques can be a source of fascination to modern deer hunters. Some tribes practiced antler rattling to entice rutting bucks; others used stuffed deer heads as puppets. Several types of deer calls were also used. Some calls were made of wood; others consisted merely of a single leaf used as a reed between the hunter's lips. Perhaps the most unusual method of luring deer was practiced by the Ojibwa (Chippewa) tribe: They often smoked herbs to attract deer.

One such herb was *Aster novae-angliae*, commonly known as New England aster. The root was dried, ground into powder, and stored in a pouch. When a deer was thought to be nearby, the hunter sat down, filled his pipe, smoked a small quantity of the herb, and waited. It was claimed that the smell was very similar to the scent of a deer's hooves.

Wildlife research biologists have proven that deer do indeed have scent glands between the toes of their hooves. These enable them to locate each other by following the faint scent deposits that are left in each hoofprint.

To Move, or Not to Move

Can't decide whether to stand-hunt or still-hunt deer? Let conditions decide for you. If it's damp, raining gently or snowing, and walking conditions are quiet, still-hunt. If the woods are dry and brittle and leaves are crackly, your chances of sneaking up on a deer are slim, so take a stand and wait for them to come to you.

Mule Deer Patterns

In a huge mountain range or expanse of desert, mule deer patterns may not be as obvious as those of whitetails in a 40-acre woodlot. But certain clues can point you toward good tree-stand sites, ambush points and stalking locations. You can pattern mule deer.

Home Range: Mulie home ranges may not seem obvious because the deer migrate from high summer ranges to low winter ranges, seemingly always moving. But in any given season, a buck will select a distinct home range and stay there. Once you find a good buck, you can bet you'll find him there consistently unless weather or an army of hunters forces him to move.

Feeding Trails: Within any home range, bucks often follow predictable paths.

Key Beds: Whenever you see a buck bedded, make note of the exact spot where he's lying. Mule deer use traditional beds day after day, year after year, and even if a specific buck gets killed, you can expect to find other bucks taking his place—and using his beds.

Escape Routes: Disturbed mule deer often follow

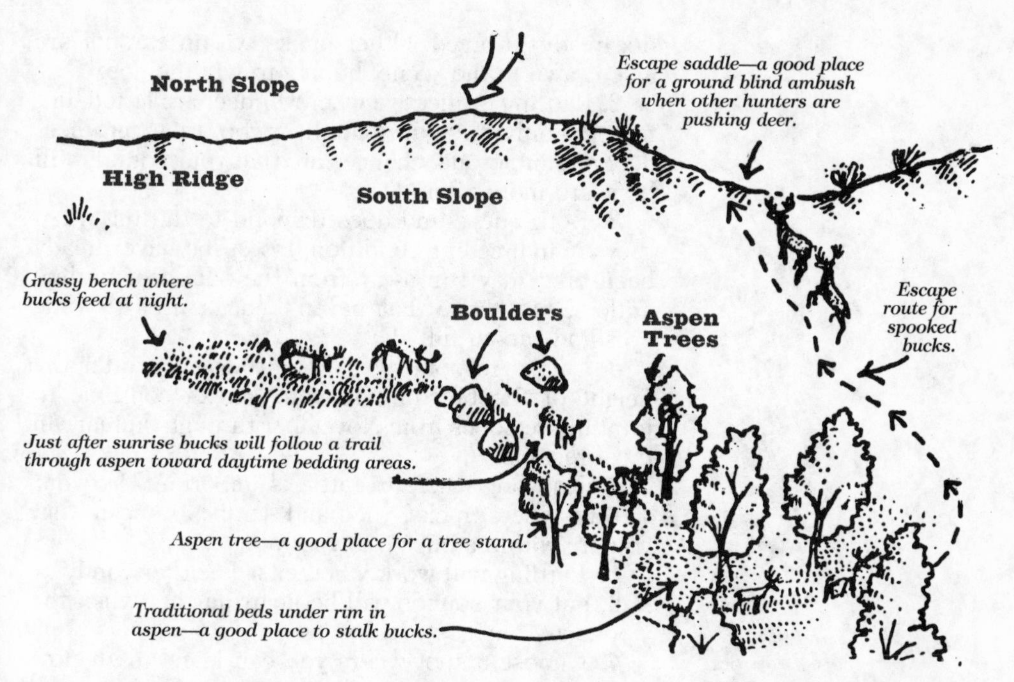
North Slope
Escape saddle—a good place for a ground blind ambush when other hunters are pushing deer.
High Ridge
South Slope
Grassy bench where bucks feed at night.
Boulders
Aspen Trees
Escape route for spooked bucks.
Just after sunrise bucks will follow a trail through aspen toward daytime bedding areas.
Aspen tree—a good place for a tree stand.
Traditional beds under rim in aspen—a good place to stalk bucks.

To trace mule deer patterns, look for feeding areas, beds, escape routes and sanctuaries.

easily identified paths of least resistance—for instance, a cut through a long rimrock, a sagebrush chute constricted by cliffs or rockslides. As long as they're unmolested there, mule deer will travel the same escape routes every time. Once they sense danger there? Forget that route. Find a new one.

Sanctuaries: Call them hunting-pressure home ranges, if you will, but when sustained hunting pressure pushes mule deer from their normal home ranges, they will seek out certain refuge areas where they are less likely to be bothered. When you come across a good-sized buck during a summer scouting trip, try pushing the deer and watch exactly where it goes. Wary bucks will head to the same sanctuary—and stay there—when hordes of hunters arrive during the season.

Twenty-three Rattling Tips

If you haven't tried rattling, you are missing one of the most rewarding experiences in deer hunting. Rattling enables you to bring deer out of hiding and lure them into close range. Here's how:

1. During the rut the sound of rattling antlers suggests that two bucks are fighting for the favors of a doe ready to breed. Other bucks within earshot are also drawn to the scene, hoping to win the doe.

2. Rattling is effective wherever deer are found, but it works only when deer can hear you. In areas where deer are numerous, chances are that your sounds will be heard more often.

3. Both bucks and does respond to rattling when they are in breeding condition. Does that have already been bred may run away from the sound of rattling antlers, but those that haven't sneak in with wide eyes and ears at full alert.

4. Bucks respond well to rattling even after the height of the breeding season. Bucks continue to hunt unbred does from November through January in most regions.

5. Knowing where to rattle is important. Do your rattling where tracks, rubs and scrapes indicate that you are within a buck's territory.

6. Rattling will work whether it is either windy or still, but your sounds will be heard farther when the woods are quiet.

7. Choose a spot where you can kneel in shadow and have a clear view in all directions.

8. Know that the deer will try to circle downwind to catch your scent as it approaches. Be ready to shoot before it reaches your wind drift.

9. Be prepared to have a buck appear within the first minute. In my experience, most deer show during the second 30-second spurt of rattling.

10. Each time you finish a rattling sequence, lay down your horns and pick up your rifle. If the deer appears while you have horns in hand, don't be afraid to reach for your gun. The deer expects to see movement. You have time to shoulder your rifle, but be sure to make your movements smooth, not jerky.

11. The deer will often appear at a distance, then put a tree between you and it and sneak in behind cover. Next thing you'll see is an antler and an eyeball peeking around a tree. To make the buck take a step into the open, brush the tips of the antlers across one another softly just once. A soft grunt call at this point often helps convince the buck to show itself.

12. When a deer approaches, it will have its eyes zeroed in on your exact location. The best time to shoulder your gun is when the deer's head disappears behind cover.

13. Heavier antlers make the loudest sounds. Some

Hands off That Scrape

During the peak of the rut, scrapes are the best indicators of buck activity in a particular hunting area. When scouting for scrapes, wear rubber-bottomed boots and don't touch the scrape: Scrapes are for deer scent, not yours.

hunters believe that such sounds scare off smaller deer and attract only mature bucks.

14. Good rattling antlers can be sawed off dead deer. Dropped antlers found in the woods may also be used if they are dry and make a sharp sound when clashed together. Don't use antlers that clunk.

15. You can keep a live sound in the antlers by applying linseed oil annually.

16. When choosing a set of rattling antlers, look for a pair with long tines and strong crotches. These will give the best variety of clashes, clicks and scraping sounds.

17. If you can't find a good set of rattling antlers, or are reluctant to cut up a trophy rack, synthetic antlers are available at some sporting goods stores.

18. Saw the brow tines off to avoid jabbing your hands when rattling, then smooth the grip area with a file.

19. Watch the terrain ahead when tracking a buck. If the trail is leading toward a patch of heavy cover, the buck may be there watching its backtrail. Before exposing yourself, stop and rattle. If the buck is there, chances are good it will come out to investigate.

20. Single bucks looking for does almost always respond to rattling. Bucks that have does with them may not.

21. When you see a deer's white flag, it has no doubt spotted your movement but probably hasn't figured out you're a human. Stop and rattle. The deer may come back.

22. When a deer "blows" at you, it is saying, "I'm a deer. What are you?" Answer by rattling. Send the message that you are two bucks fighting, and the deer may come back to investigate.

23. The best and safest way to haul antlers is in a blaze-orange daypack.

✳ ***How to Rattle:*** Begin with a solid clash of the antlers, then twist them quickly back and forth in an interlocked position to simulate the sounds of sparring bucks. Rattle this way for approximately 30 seconds, then separate the antlers by twisting them under tension and pulling them apart. Wait one minute. Then follow with a repeat performance. Continue to produce 30-second spurts of rattling sounds every minute or so for about 10 minutes. If you don't see anything, move ahead several hundred yards and repeat.

Glass for Thick-Timber Bucks

Whitetail habitat is generally dense, and because of this, many hunters may underestimate the importance of glassing with binoculars. For one thing, glassing will help you separate a deer from the foliage: An eye, ear or antler that might have gone unseen will stand out at 7X. Early mornings and late afternoons are the best times. A hunter at the edge of timber can glass large open fields for feeding deer (which are likely to remain in the field for a time) and then plan a stalk.

Blacktail Tactics

If you're like me and your hunting time is limited, go out late in the season and hunt early and late in the day. Try to locate inaccessible areas where the average hunter won't take the time to hunt. Depending on terrain, these areas should be a half-mile to a mile away from any roads or heavily hunted areas. Look for secluded hillsides, near good food sources. Look for clear-cuts or burns from three to five years old—i.e., logging operations that have scattered open areas still remaining. Hunt on the edges of this habitat, where the open areas merge with heavy timber or brush-covered hillsides or draws. Blacktails are fringe animals, spending most of their time in or near these locations.

The most effective manner of hunting this type of country is a combination of still-hunting and stand-hunting. Slowly and carefully still-hunt through your chosen area, and when you see an especially good site, such as a series of heavily used intersecting trails or a spot that just looks enticing, sit down and watch for a while. If nothing shows, continue to still-hunt until you find another spot, and so on.

The best still-hunting conditions are a light, misty rain (with the wind blowing in a steady direction) or right after a heavy rain that has continued for most of the day and stops an hour or two before legal shooting light. The light rain softens noisy leaves and branches and causes a steady drip from vegetation, which helps to camouflage human sound. Also, the grayness caused by the low cloud cover causes deer to head for bedding later in the morning and to come out to feed earlier in the evening. Heavy rains and gusty, unstable wind conditions tend to cause deer to be very jumpy and nervous. This makes effective still-hunting much more difficult.

Use your binoculars often and analyze every piece of cover. Bucks don't grow large antlers by standing in conspicuous places. In this brushy country, a wide field of view is not important. Remember, the idea is not to see how much landscape you can see or cover, but rather how much you can see in the landscape.

[Blacktails are found in California, Washington and Oregon. The Sitka blacktail now has its own designation for Boone and Crockett purposes and is found in certain parts of Canada and Alaska.]

Hunting the Off-Hours

During the primary rut, try hunting off-hours from 10:00 a.m. to 2:00 p.m. This tactic works for a couple of reasons. First, you'll be hunting at a time of day when fewer hunters are in the woods. Second, because deer are less pressured at this time, they move about more freely, giving you the chance to see more of them.

STEVENS

ODORLESS GUN OIL

OUR LATEST FIREARM ACCESSORY

**"Best by test—
Superior to all the rest."**

This Oil is a lubricant, rust-preventiye, polishing and cleaning compound. It is guaranteed to be absolutely pure and to contain no acid. Especially adapted for Firearms, Bicycles, Sewing Machines and all Mechanisms requiring a High-grade oil.

**Price in 1 ounce bottle, 10 cts.
In 4 ounce bottle, - 25 cts.**

OUR LINE:

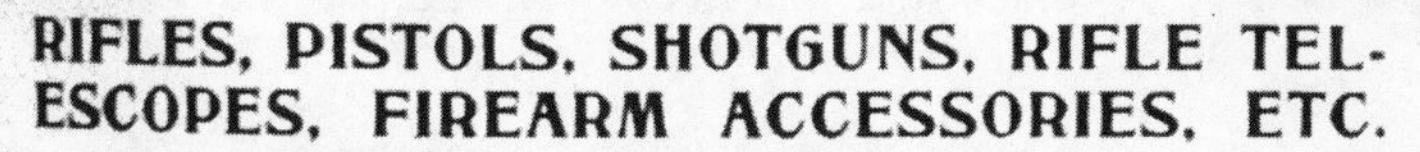

RIFLES, PISTOLS, SHOTGUNS, RIFLE TELESCOPES, FIREARM ACCESSORIES, ETC.

Your local merchant should handle the STEVENS. Ask him. If you cannot obtain our popular models, we ship direct, *express prepaid*, upon receipt of Catalog Price.

Send 5 cents in stamps for 160-Page Illustrated Catalog. An invaluable Manual of Reference for all who shoot, or are going to.

Ten Color Lithographed Hanger mailed for Six Cents in stamps.

J. STEVENS ARMS AND TOOL COMPANY,

P. O. Box 5680,

CHICOPEE FALLS, MASS., U. S. A.

A Quiet Deer Drive

Deer hunters are taking a hard look at noisy drives in which large numbers of hunters crash through the woods with abandon. Hunting like this is often inefficient and unproductive, not to mention disruptive.

A successful deer drive should be shrouded in stealth and aimed at placing many hunters in optimum positions. The drive territory should include several kinds of cover: open woods, small areas of security cover, and deer trails from each dense bit of brush where deer might bed down for the day. The group should include drivers and standers in a proportion that fits the size of the territory. The drivers should advance into the wind, and the standers should be waiting downwind. All hunters who will drive and stand should be stationed with the utmost caution. Noise is an absolute taboo until the drive is ready to begin.

The pace of the drive can never be too slow. It's quiet hunting all the way: moving snail-slow, scanning the area, staying on game paths, walking, waiting, and listening. Be alert for deer sneaking around and behind, and watch trails that intersect. Deer moved by the drivers will run nose to the wind, in the direction of the hunters on stand.

The hunters on stand must be aware that trails leading in will be used by drivers as well as deer, and they must take a position where there is no chance of shooting back into the brush. They should instead place their shots carefully across the trails.

Would You Be Quiet?

During dry seasons, when walking in deciduous forests proves noisy, try still-hunting in young pine forests. Big bucks don't like to walk through crackly woods any more than a hunter does. If hardwoods are dry and noisy, they'll often move to quieter coniferous forests.

ITHACA GUNS.

Self-Compensating, taking up wear at every point.

Guaranteed not to shoot loose with Nitro Powder. Close, Hard-Shooting Guns a Specialty.

Prices Reduced and every Gun warranted in the most positive terms, Send for Circulars and ask for special prices.

ITHACA GUN CO., - ITHACA, N. Y.

5

Bows, Arrows and Blackpowder

Yes! let us take down the old rifle and be always among the trees. Let us seek contentment in the seclusion of the forest and let us ever keep alive in us the old hunting instinct, which in youth would rise up in our hearts at the first whiff of October air; and with rifle on shoulder and powder horn hanging at our side, let us seek again the old perfect life.

—Paul Griswold Huston, *Sports Afield*, November 1899.

A Bowhunter's Whitetail Checklist

Some hunters like huge stands complete with a buffet of munchies, a stocked library and a full wardrobe for comfort under any condition; others prefer more spartan trappings, taking only lightweight clothes and a tiny portable tree stand.

Realistically, one extreme is just as bad as the other. Yes, to hunt long and well you have to be comfortable, so a roomy stand with a seat is functional. But take a load of gadgets and goodies and they'll only get in your way, reducing your efficiency. Try to compromise by picking all your gear according to one criterion: function. Every item must promote hunting efficiency. The list here includes all the standards you should take on an average day of whitetail hunting. These are not the best for everyone, but perhaps this will give you some ideas for your own gear list:

Clothing:

✷ Camouflage fleece or wool outerwear
✷ Long johns (medium Thermax or polypropylene)
✷ Down or acrylic-pile vest
✷ Scent-Lok suit
✷ Knit hat with earflaps
✷ Wool gloves
✷ Hand muff
✷ Knee-high, insulated rubber boots
✷ Rain parka
✷ Camo facepaint (or headnet)

Archery Tackle:

✷ Bow (equipped with sight, string peep, overdraw, shoot-through arrow rest, eight-arrow quiver)
✷ Arrows (all broadheads, except two arrows tipped with bludgeon heads for practice)
✷ Release aid (wrist strap, caliper jaws)
✷ Armguard (to keep coatsleeve out of bowstring)
✷ Binoculars (6X, lightweight)
✷ Rangefinder
✷ Trail-marking reflectors

Hunting Gear:

✷ Portable tree stand (in some cases it's best to use a fixed-position stand and 10 strap-on tree steps; in areas

with straight-trunked trees, use a lightweight climber)
* Climbing and safety belt
* Bow hanger
* Tree seat

Fanny pack for carrying the following small items:
* Limb saw
* Pruning shears
* 20-foot pull-up rope
* Flashlight
* Compass
* Knife
* Calls (grunt, doe bleat, rattling antlers)
* Scent-eliminator spray
* Paper towels
* Pack frame or deer drag (depending on location)
* Water
* High-energy foods
* Notebook, pencil
* Camera

How to Sight In a Bow

Some bowhunters seem to have more difficulty sighting in their bows than is necessary. Before you begin to sight it in, your bow must first be in tune in order to achieve good arrow flight.

Begin with the nock-set. The plastic-lined metal type works well, especially if you use two of them butted tight against each other. Locate them so that the bottom one is three-eighths of an inch above center on a bow square. They should be crimped on with a nock-set tool (regular pliers will only ruin them). If you don't have a bow square or a pair of nock-set pliers, visit a pro shop.

Assuming that your arrows are matched to your bow in length and spine, you can begin. (A cushion plunger will make the job easier and improve your arrow flight.) Start with the least amount of tension the plunger will allow.

Now nock an arrow and look down the shaft, holding the bow in your left hand, as if you were about to take a shot. Keep both eyes open and line the string up with the center of the shaft. Screw your sight pin out until it rests on the same plane as the shaft and bowstring. That approximates your windage.

Start shooting at five yards on a target that is big enough not to be missed. The arrow should impact to

How to Ship Your Bow on a Plane

When big-game hunters fly on commercial air carriers, they commonly use rectangular hard-sided guncases. Moreover, the airline baggage clerk always affixes a bright red tag to the case proclaiming an unloaded firearm inside. This causes untold grief to bowhunters who commonly use the same type of case. Solution: When checking a hard-sided case containing archery equipment, always have the baggage clerk attach the same type of red tag but with the word "firearm" crossed out and "no gun inside" written in its place.

the right of your point of aim. Increase tension on the plunger button until your shots come to exact center.

Now make your elevation adjustments, moving the pin down to raise impact and up to lower it. Once you have this pin set, you can line all others up with it for windage and then set the elevations individually for whatever distances you desire.

Experienced bowhunters usually sight in to their own paced distances rather than set yardages, so they can pace off landmarks in unfamiliar terrain.

Bowshooting Basics

S*tance:* A study of Olympic-class archers revealed leg strength as the major variable in predicting success; archers with the strongest legs had the highest scores. That's because strong legs give the upper body a solid foundation. But that foundation also demands the proper stance. To develop a solid stance, place your feet shoulder-width apart, 90 degrees from the target, and then step back a half step with the front foot and pivot slightly toward the target for a somewhat open stance. Keep your weight evenly centered on both feet, and stand up straight. Don't lean forward or back. Your shooting stance should remain completely unchanged throughout the shot.

✳ ***Bow Hand:*** The meaty part of your thumb should be the center of pressure on the bow handle. Don't palm the bow; you want only one pressure point. To ensure that, do not hold your hand vertically, but rotate it slightly in the same natural position you would when pointing your index finger. With your hand in that relaxed position, your little finger will hang to the side of the bow handle, not in front. Also, the fingers on your bow hand should be relaxed, hanging loosely throughout the shot.

✳ ***String Hand:*** First, hook your fingers on the string at the first joint, and then, like the bow hand, rotate this hand slightly into a natural position. If you hold your hand straight up and down, it will try to rotate back to a natural position as you draw, twisting the string.

Second, keep your wrist straight, the most relaxed position. Your fingers should simply be a hook on the end of a straight rod.

Third, center bowstring pressure on the middle finger. Mechanical release aids work well because

Arrow-Spin Control

Many arrows are made with the vanes straight. That type of fletch is fine if you are shooting field points, but once you replace them with your broadheads, arrow flight becomes erratic. The solution is a helical fletched arrow, one with curved vanes, or feathers, that impart a spin to the arrow. It's much the same as rifling in a gun barrel. An archery pro shop can change the vanes on your arrows and adjust the degree of curvature to suit your equipment. The right amount would be just enough to stabilize the broadhead; too much will slow it down.

All serious bowhunters should practice shooting on stand.

they have a tiny point of contact on the string; you want to duplicate that with your fingers. The middle finger should hold most of the weight while the other two fingers "float" on the string.

✱ ***Bow Arm:*** A solid bow arm is critical because it holds your sights on target. Even with a bad release, you'll shoot accurately if your bow arm is solid. But ironically, that means it must be relaxed. Think of your bow arm as a brace. Once you've raised your bow into shooting position, the arm locks into place, and you can relax the muscles.

Your bow hand can be a valuable checkpoint here. Again, it must be relaxed. A choking grip on the bow handle, or stiff fingers like wheel spokes, indicate tension in your arm. Relax your bow hand and arm. Also, do not hunch your bow shoulder; instead, pull it down to form a straight line from your wrist to your neck. That's the most stable position.

✱ ***Release:*** To release smoothly, never open your fingers and let go of the string. Rather, gradually relax your string hand and fingers until the string slips away. Likewise, with a release aid, don't jerk the trig-

Practice Daily

Ten or 15 minutes of practice with your bow every day or so is better than a marathon session once a week. Concentrate on every shot and imagine you are pulling back on a huge buck.

ger, but slowly squeeze until the aid goes off. With both methods, you should be surprised when the bow goes off.

After the shot, analyze your string hand. It should move straight back along your face, and the fingers should be limp. If your fingers are stiff or your hand has moved forward or out from your face, you've done something wrong. Concentrate on relaxing your hand and let the shot surprise you.

✳ ***Follow-Through:*** After the shot, continue to aim at the target. If your bow jerks one way or another, you've shot with tension. Try adjusting your stance, and analyze your bow hand, bow arm and string arm to eliminate tension that could be torquing the bow. As professional archer Scott Wilson said, the release should be only a momentary interruption of the aiming process. That's follow-through.

Bowhunting Terminology

Like most sports, bowhunting has a dictionary of terms all its own. Here are the most common ones, with brief definitions:

✳ ***Broadhead:*** The "bullet" part of your arrow, the penetrating, cutting blades. Made of steel, these razorlike edges are available in a wide variety of models, although most are essentially similar, with three or four blades.

✳ ***Cams:*** The wheels located at the tip of each limb of a compound bow that help determine the percentage of let-off when the bowstring is drawn.

✳ ***Compound Bow:*** A bow with cams that features a let-off, or reduction in the pull needed to draw a bow.

✳ ***Draw Length:*** The distance, measured in inches, that you pull a bowstring back. More specifically, it is

Recurve-Bow Components

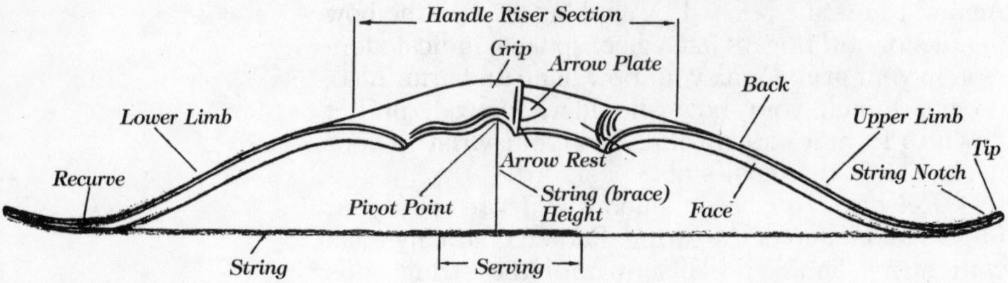

Handle Riser Section
Grip
Arrow Plate
Back
Lower Limb
Upper Limb
Tip
Arrow Rest
String Notch
Recurve
String (brace) Height
Pivot Point
Face
String
Serving

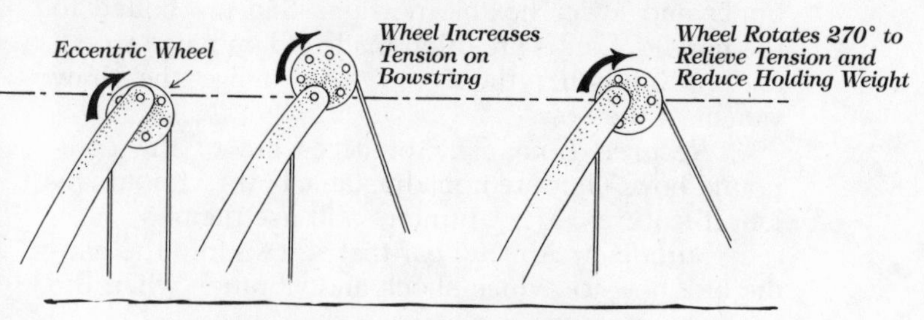

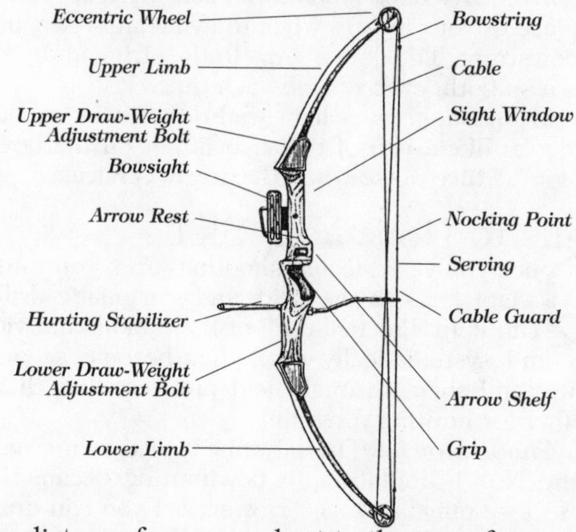

the distance from your chest to the tips of your middle fingers when your arms are extended. This measurement is essential in fitting a bow properly to the shooter.

✱ ***Draw Weight:*** The amount of pull needed to bring a bowstring back into shooting position. It is measured in pounds; most quality compound bows have an adjustable draw weight.

✱ ***Fletching:*** The plastic or feather material on the rear of the arrow that stabilizes it in flight.

✱ ***Jumping the String:*** The amazing ability of a white-tailed deer to react to string noise and actually dodge an arrow in flight. To minimize this problem, try to shoot when the deer's head is down, or use string silencers.

✱ ***Let-Off:*** The reduction in the amount of pull needed to bring a compound bow to full draw. Many

String silencer

compound bows have a let-off of 45 to 50 percent.

✱ ***Limbs:*** The "arms" of a compound bow, the upper and lower flexible portions that are bolted to the handle. Limbs are often made of laminated wood, and by adjusting them you can change the draw weight of the bow.

✱ ***Recurve Bow:*** The predecessors of the compound bow. They are nonadjustable bows without any let-off features. Many hunters still use them.

✱ ***Stabilizer:*** A metal bar that screws into the handle of a bow to reduce shock and vibration when the arrow is released.

✱ ***String Release:*** This mechanical "trigger" takes the place of your fingers when drawing and releasing the bowstring. The device clips to the string and provides a smooth, even release each time.

✱ ***String Silencer:*** A light wad of nylon or rubber that looks like a ball of twine. It is put on the bowstring to reduce noise when the arrow is released.

Drills to Hone Your Skills

Once you've perfected shooting form, you must adapt form to the field, and systematic drills can help. Practice each of these skills individually and systematically until they become second nature, and then, through field practice, link them together for hunting versatility.

1. Shoot Slowly: The movement of drawing may be the No. 1 limitation in bowhunting because it alerts close-range animals. However, if you can draw so slowly that an animal, even looking at you, fails to see the movement, you'll lose few shot opportunities. With the jerky draw cycle of a compound bow, developing a "motionless" draw isn't easy. To perfect it, hold your bow in shooting position, aim and draw as slowly as possible. Your sights should hold steady on the target, and your string hand should come back steadily with no jerks or pauses. Do this six to 10 times per session, building up to a full 10 seconds per draw. If you can't draw your bow straight back, the draw weight is too heavy.

2. Shoot Fast: At times you must be able to draw and shoot quickly. Some archers would say a stick bow and an instinctive style are fastest, but with practice you can shoot nearly as fast—and more accurately—with a compound bow. To develop efficiency, see how many arrows you can shoot in a one-minute period. Then see how quickly you can extract and shoot all of the arrows from your quiver. These are great drills for shooting efficiency. But remember, all shots must be accurate; wild shots mean nothing.

3. Shoot Long: Follow-through, the ability to hold your bow in shooting position after release, is central to accuracy, and there's no better way to practice follow-through than long-range shooting. If you're standing 70 yards (210 feet) from the target, and your bow shoots 210 feet per second, the arrow takes one second to reach the target. That's a long time, and you'll want to drop your bow arm to watch the arrow. Don't do it. Discipline yourself to hold your sights on the target until the arrow hits. Not only will you develop a solid follow-through, essential at all distances, but you'll build confidence at short ranges. Do not take long-range shots on animals—30 to 40 yards is still the maximum practical range.

4. Shoot in Bad Weather: To prepare for actual hunting conditions, practice in wind, rain and other bad weather. By so doing, you not only learn to shoot well under adversity, but you discover essential details about your tackle. In wet snow you might find that your arrows and rest ice up; in rain, you might find that water plugs your peep sight or makes your cable slide squeak. Only by shooting in different types of weather conditions can you analyze and correct subtle problems.

5. Shoot in All Positions: In hunting, you can't always adopt an ideal shooting stance. You may have to shoot around trees, under limbs, or straight up or down hill. Systematic practice will prepare you for all contingencies. To start, practice regularly while kneeling, sitting, leaning to the side, and at steep angles up and down. You may find you simply can't shoot accurately from some positions. That's good to know. Eliminate those and develop positions that work best for you. Then try roving around your range, quickly

assuming and shooting from many different positions.

6. Shoot Blind: Fumbling with a release aid or other gear has cost many archers good shots. Practice can prevent that. If you shoot with a release aid, for example, practice nocking arrows and clipping the release onto the string with your eyes closed or in the dark when you can't see. If you can perform all shooting functions by feel, they'll come smoothly and automatically in the field. (Incidentally, if you shoot with a release aid, learn to shoot the same bow with fingers. It requires some sighting and anchoring adjustment, but it gives you a backup if you lose your release.)

Dry Fletchings

A simple and effective way for bowhunters to keep their fletchings dry on a wet day is to cover them with a plastic bag, such as the bags you are given at the supermarket. These light, waterproof bags can be fitted over the end of an arrow, twisted to stay in place, and yet are instantly removable for a shot.

Check Bowstring Length

In addition to routine bowstring care and inspection, the string should be carefully checked for overall length to forestall accuracy problems. While some string materials resist stretching better than others, none are immune. This is particularly

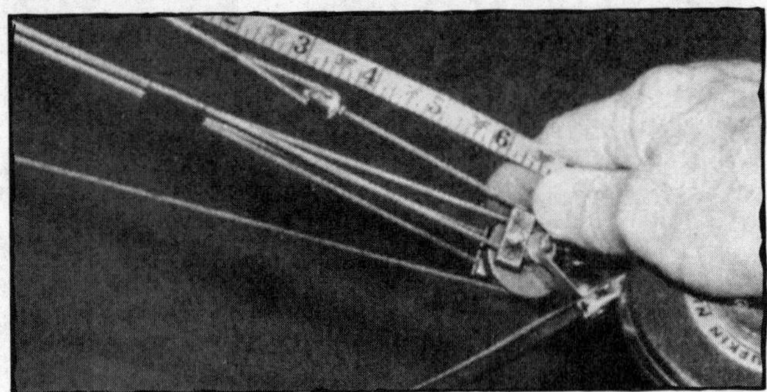

For consistent accuracy, a bowstring needs adjustments.

true of newly installed Dacron strings on bows of heavy pull-weight. String life of heavier bows is also calculably shorter.

To measure overall string length, carefully select a reference point at each end of the string and, using a flexible steel tape, measure to the nearest ¼ inch. Strings that vary more than ¼ inch between measuring intervals are likely to show noticeable accuracy and tuning changes. Applying torque to the string will shorten it. Carefully note the number of turns required to restore to normal length and keep a record of this.

Each bow is likely to have a slightly different overall string length unique to its mechanical and structural makeup. Periodic inspection and adjustment will be necessary during its shooting life unless major components are replaced.

Makeup, Please

If you're a bowhunter who applies face camouflage, take a few tips from Broadway makeup artists:

Strong, overhead stage lights (which are similar to sunlight along a field edge) exaggerate facial features. Eye sockets darken and loom large. Cheekbones flare. Nose and chin jut, which is fine if you're playing Dracula. To overcome adverse lighting, makeup artists countershade an actor's face. You can do the same. Why look like a monster to a monster buck?

To make your face look less facelike to a deer, smear dark camo paste on normally light areas: forehead, ridge of nose, flat of chin, cheekbones. Smear light camo paste on normally shadowed areas—under eyes, on bridge of nose, beneath nose, beneath lower lip, under jaw and chin, in hollows of cheeks, along eyebrows.

Now, on a bright day in the woods, you'll be able to disappear.

Label Your Bowsight Pins

With the advent of programmed cams and lighter, shorter arrows made possible through the use of overdraws, bowhunters are currently reaping the benefits of a much-improved trajectory. Greater arrow speed does pose at least one problem for the sight shooter: The flattened trajectory brings sight pins so close together that the chance of using the wrong pin becomes a distinct possibility, especially in critical hunting situations. The best way to avoid making this mistake is to label the pin for its set yardage, so the label can be seen in the aiming point area.

To do this, type distances on a self-adhesive label, like those used for files or cassette tapes. Then cut the label just slightly wider than the typed number and about a quarter of an inch long. To see the number on the sight pin, first build up the area of the pin where the number will be placed using a thin strip of masking tape. This location should be away from the aiming point of the sight pin so as not to restrict light, yet close enough to be instantly visible during normal

A Bowman's Clothes and Shoes

Clothes must be soft and nonrestrictive, which eliminates heavy cotton (including jeans) and nylon. These have no place in stalking. Instead wear soft net clothing or lightweight (two-ounce) cotton in warm weather (no jeans or other binding clothes underneath); Polarfleece, knit acrylic or wool in cool weather. These whisper-soft materials allow you to move quietly and freely.

Wear soft, pliable shoes. Hard lug soles may be fine for packing heavy loads, but they're terrible for stalking. In warm weather, try lightweight running-style shoes. In noisy conditions, you'll do even better in stocking feet. In damp or cold weather, use rubber-bottom, leather-top boots.

sighting. The typed numerals are then simply applied to the masking tape on the pin, facing the shooter.

Naturally, the pins should all be sighted in before they are to be labeled, as the labels will restrict windage adjustments if any have to be made afterward, and the labels may not realign properly should any windage change be made.

To weatherproof my yardage indicators, I cover them with a strip of transparent tape. At full draw, even through a string peep, the numbered pins show up well in all but the poorest light.

Make Your Own Bow Hanger

To make a light, handy bow hanger, use a television wire standoff screw. Wrap the looped end with electrician's tape to protect the bow limb and prevent a scraping noise when you lift off the bow. Screw this into a limb or the tree trunk well above your head (be sure to remove it when you leave) so that when you see a deer approaching, all you have to do is slowly reach up and bring it down. The arrow should already be in position on the string in an arrow holder.

Reference Arrow Helps to Retune

Marking an arrow shaft at full-draw position can provide helpful information later. Bows often become misadjusted for no apparent reason, and a change in draw length can be a clue. String length, cable adjustment or other visible mechanical changes can alter the full-draw position of your bow. Measuring and recording critical setting will save you time and effort whenever you need to retune or check your bow.

A discarded arrow shaft serves nicely as a marker arrow. At full draw, have someone put a mark on the arrow at an exact point on the bow frame or arrow rest. Do this for each bow you own, and put the arrows away for future reference. Then, at intervals, or when your bow shows changes in accuracy or tuning, use the arrow to check draw length.

Unorthodox Tips for the Bowhunter

1. Just about the best scent you can use to lure deer is sardines. Well within bow range, open a can of them and place it under a small bush. A buck running full tilt will put on the brakes to stop and take a whiff, almost every time.

2. A hot autumn will put the stopper on the rutting season, but it won't slow down the deer flies all that much. Spend the early morning hours on the stand; after the temperatures climb, still-hunt the swamps. Deer lie down in these places to keep cool, and they will be swinging their heads to dislodge deer flies. Keep your eyes peeled for movement: You move when the deer is concentrating on flies.

3. A small transistor radio can help you call in many a deer. Just get on your stand, turn on the radio to a station you like, with the volume low, and wait. Watch closely; they will come like a puff of smoke, quietly and slowly. Just don't play loud rock 'n' roll!

Don't Take That Head Shot

While a brain bow shot on a deer can result in a quick kill, the odds are very much against it. Here's why: The fist-sized brain of a mature deer is encased within a surprisingly well-protected cranial vault. On a buck the entire head will be protected by the antler except on a direct frontal shot. A brain shot requires great accuracy on a very small target. Even seemingly well-directed broadheads can easily glance off the thick skull or penetrate at an angle, resulting in a nasty, crippling wound.

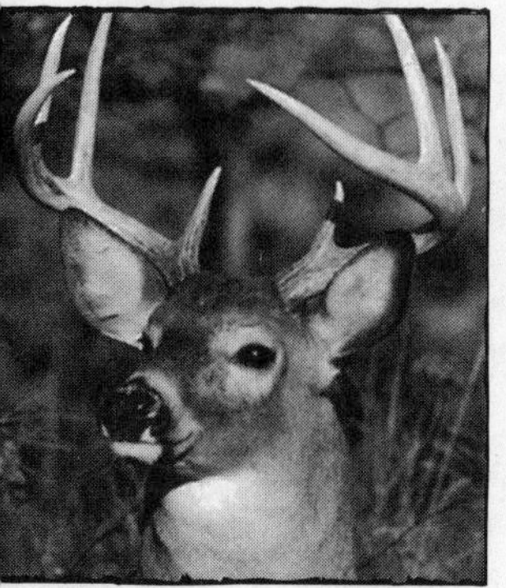

For the bowhunter, a head shot is risky at best.

A faster-than-average arrow velocity from a heavy arrow would be a minimum requirement. An arrow velocity of at least 200 fps is recommended, along with a very durable broadhead. However, the most well-constructed broadheads may have their blades stripped or shattered on impact with a deer's skull, thus causing only partial penetration. A poorly directed arrow may also hit adjacent nonvital areas of the deer's head, not killing it but causing incapacitation or a slow and painful death.

Waterproof Bowhunting Gear

Bowhunters fussy about moisture on their firearms often neglect their bowhunting equipment in identical situations. Just because bows and arrows are often made of wood, fiberglass and aluminum does not mean that they can be schlepped around in rain without harm.

Moving parts (eccentric wheels, cams, idlers, etc.) require lubrication to protect against wear and corrosion. Another component that requires attention is the bowstring. Usually made of Dacron, it will absorb water unless protected, which could result in slow and slower shots. A liberal coating of bowstring wax

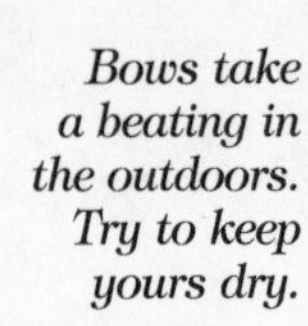

Bows take a beating in the outdoors. Try to keep yours dry.

burnished with a piece of leather will do the job.

If the shaft of your arrows are fletched with feathers instead of plastic vanes, the feathers will also pick up moisture and slow the arrow down. They can be protected by applying dry-fly silicone spray.

Tree-Stand Tips

Both tree stands and ground-level blinds have their good and bad points. Most bowhunters, however, prefer to hunt whitetails from above. Here are some of the reasons:

While it's true that a deer can look up and see you when you're in a tree, it generally isn't looking for you there. Even when it does spot you—usually because you made some slight noise or movement at the wrong time—there's no way it can get to you to satisfy its curiosity.

With a ground blind it's a different story. All the deer has to do is circle downward until it catches your scent, or simply walk up and peer in at you. In either case, you aren't likely to get a chance for a shot since the deer is alerted and looking for an excuse to run. From a tree, though, after the animal has tired of staring at you, you can often get a shot as it moves off.

Some hunters think that when they're up in a tree the deer can't smell them. This is only partly true. A deer will spook as far as 100 yards downwind of your tree stand. However, there is a magic circle of about 30 yards or so around the tree where, most of the time, your scent will be over the deer's head. This gives you many options when picking stand sites that don't exist when hunting from the ground.

Hunting from a tree does have some drawbacks. The obvious one is that often there won't be a tree where you want to set up. Also, if a deer walks by just out of range, you can't simply pick up and go after it or try to intercept it down the trail, as you might on the ground.

Ten Tips for Stalking Mulies

In bowhunting, you must see animals before they see (or hear or smell) you. That's why stalking works so well. It's based on the idea that you locate animals at a distance before they detect you.

With an undisturbed animal located, you're 90 percent home-free. You can stalk big game in all kinds of terrain and cover, too. I've stalked deer in the wide-open West, and in tough, dense country. It's just a matter of finding pockets where you can see, even if only 100 yards.

1. Spot Your Deer First: Spotting is especially on a mule-deer hunt. Use high-quality, full-size binoculars. For broken or brushy country, use 6X to 8X; for alpine, prairie or desert terrain, go with 8X to 10X. In the West, add a spotting scope of 20X or 25X, or binoculars of 15X to 20X. Any optics over 10X must be tripod-mounted. You can also locate animals, particularly elk, by calling, and stalk them by sound rather than sight.

2. Plan Your Stalk: Once you've located a deer, take time to plan your route. Note landmarks around the animal to help guide you to him, and try to locate all surrounding animals that might get in your way and spook the deer you're after.

3. Ambush Moving Deer: Rarely can you keep up with feeding animals without being seen or heard, so rather than trying to move straight in, anticipate their route, circle ahead, and wait for them to come to you. The ambush approach is particularly useful on noisy footing, such as shale or crusted snow.

4. Wait for Deer to Bed: In open country or cliffs, where you can't move quickly without being seen or heard, wait for animals to bed for the day and then stalk them. That way you have several midday hours to sneak within range. Mule deer will stay put from about 10:00 a.m. to 4:00 p.m.

5. Get the Wind: Don't expect miracles from deodorants and cover-ups. Stalking is sweaty business, and your only hope is a favorable wind. Frequently use a squeeze bottle filled with cornstarch, or a butane lighter, as a wind indicator; if the wind isn't right, back off and try later.

6. Move Slowly: To keep from being seen or heard, picture yourself as

Tools for Tracking Game

These aids can help you recover animals:
1. String tracker to keep you on tough trails.
2. Plastic flagging to mark trails.
3. Binoculars to spot animals far ahead and in thick brush.
4. Flashlight or lantern to aid in night trailing.
5. Compass to systematize search patterns.
6. Sure Sign and similar products to make blood more visible.

the hands on a clock, making progress without visible movement. If it takes you an hour to cover 100 yards, so be it. Close-range stalking success can be summarized in two words: slow movement.

7. Hide: Use boulders, draws, cliffs or other solid cover for concealment. If necessary, crawl to stay below the brushline. Wear clothing that blends with surroundings—tan, gray, light green in the desert; charcoal, brown, black, dark green in the woods; white on snow. Camouflage your face and hands in similar shades.

8. Use Your Binoculars: Relocating a buck once you've changed position can be the hardest part of a stalk. Keep your binoculars around your neck (use an elastic band to hold them tight against your chest), and use them constantly as you get close to look for an antler tip, twitching ear or black nose.

9. Use a Rangefinder: In stalking, a rangefinder really works because you can usually see some small part of the animal, like an antler tip, to range on. Knowing the exact range means clean kills (this assumes you use bowsights.)

10. Wait for the Sure Shot: You can't always expect to stalk within tree-stand ranges. Know your effective range, and whether it's 30, 40 or 50 yards, once you get within that range, stop. Calm, unaware animals present the best shots, and the worst possible mistake is trying to get too close and alerting the animal. Stay back and wait for it to move into good shooting position.

Broadhead Points

The broadhead may be the determining factor in whether you make a kill or not. Weight difference in the nose of the broadhead can cause the arrow to fall short or carry over the target. The size and style of blades affects the penetration and the bleeding of the animal. The greater the number of blades, the more meat is affected. It is important to match the broadhead to the conditions you're facin

What's Stalkable With a Bow?

Stalking works on all big game under the right conditions, but some animals are more stalkable than others. Pope and Young records show the following percentages for species taken by stalking:

Species	Percent	Species	Percent
Mountain Goat	*90*	*Moose*	*60*
Bighorn Sheep	*85*	*Antelope*	*30*
Caribou	*75*	*Elk*	*25*
Mule Deer	*70*	*Black Bear*	*10*
Sitka Blacktail	*65*	*Whitetail*	*5*
Columbia Blacktail	*60*		

Arrow Rests: Why Details Matter

An arrow rest appears small and insignificant, but it may be the most important part of your bow because it's the last point of contact before your arrow launches into flight. How well the rest functions determines how well the arrow flies.

Functions:

1. The most basic is to hold your arrow until you shoot. No big deal? If the rest breaks or loses adjustment, your hunt could be ruined.

2. The rest controls lateral adjustment of your arrow—or centershot, as many call it—the very basis for bow tuning.

3. To some degree, the rest affects arrow spine. Lateral adjustment and spring tension in the cushion plunger (if you use one) in essence change spine value.

4. Fletching contact, especially with vanes, is a major reason for bad arrow flight; the arrow rest has a major effect on this contact.

Qualities:

1. Durable: Extreme cold can harden the plastic backing, which will snap at the bow's vibration.

2. Secure: Screws must lock into place so the vibration of shooting doesn't loosen adjustment.

3. Quiet: There can be no hissing or squeaking as you draw your bow. Not important? On a quiet morning, try drawing on a deer or bear 10 yards away.

4. Easy to Adjust: Some rests are complicated, some simple. Choose the simplest for your purposes.

Styles:

1. Launcher: The arrow is set on a V-shaped platform and the fletching passes on both sides of the platform. In essence, the arrow is balanced on the platform and can easily fall off. For this reason, launchers are best for target shooting, not hunting.

2. Shoot-Through: The arrow is cradled securely between either two support arms or an arm and a cushion plunger. One vane of the arrow passes between (shoots through) the two parts of the rest. This is an excellent hunting style for use with a release aid, which lets the arrow glide straight through the rest. Fingers cause an arrow to flex more, and the fletching will probable hit this type of rest.

3. Shootaround: These consist of a movable arm

Eliminate Bow Noise

When an arrow is released, the hollow sockets in a compound bow can amplify the noise of the vibrating limbs. To muffle the sound, back the limbs out of the sockets, and fill the space with insulating foam or caulk. After the material is dry, screw the limbs back into position. If you take them clear off to fill the cavity, screw the bolts back in before filling to keep the female threads free of caulk and maintain the hole for the bolts.

Silence the rattling arrows in your quiver with styrofoam.

and either a semi-rigid sideplate or a cushion plunger. Under this general heading are: a) solid plastic rests; b) the springy rest, a coiled spring with a long tail to hold the arrow; and c) the flipper, a spring-loaded or magnetic arm that flips inward as the arrow passes by and then flips back to its original position.

Of these three, the flipper, with a metal sideplate or cushion plunger, is probably the best. It's adjustable and assures the best clearance.

Draw-Weight Dilemmas

Many hunters shoot bows of 70 to 80 pounds, and others feel they have to keep up. How much draw weight do you need? Any modern bow of 60 pounds or heavier, well tuned and shooting sharp broadheads, will assure more than adequate penetration for animals up to the size of elk.

So what's the advantage of heavier weights? Trajectory. Draw weight is a—if not the—major influence on arrow speed. The heavier your bow, the faster the arrows will travel, the flatter the trajectory and, ultimately, the greater your accuracy, especially at unknown distances.

That sounds ideal, but at some point you may lose more accuracy by increasing weight than you gain by flattening trajectory. Accurate shooting comes from relaxed form, and if you have to strain to pull a bow, tense muscles can lead to erratic form or, worse, target panic. Also, many hunters, fatigued from hard hunting, are unable to draw their bows when they finally get a hard-earned shot.

Bow control is the real key to accuracy, and that means a comfortable draw weight. First, can you hold a bow at arm's length and draw it straight back smoothly? If you have to raise the bow and lunge to break it over, it's too heavy. Dave Holt, author of *Balanced Bowhunting*, suggests sitting on the ground and drawing the bow. If you can't do that, the draw weight is too heavy for you.

How to Aim More Quickly

Shooting with sights can be faster than without because the sight assures a positive aiming point. These tips can increase aiming speed:

1. Even with sights and string peep, always shoot with both eyes open.

2. Once you're on target, begin to execute the shot. Don't hold and hold, trying to steady your bow; you'll only wobble more.

3. Don't estimate distances to the precise yard; estimate and aim in five-yard increments. That is, for a target between 23 and 27 yards, hold for 25; for a target from 28 to 32 yards, aim for 30.

Tune Up Your Shot With a Chronograph

Like a smooth-running car, an efficient bow always gives the best overall performance, meaning consistent accuracy and good penetration. The easiest way to gauge a bow's efficiency is to measure arrow speed, because, for any bow at a given draw and arrow weight, the most efficient setup will shoot the fastest.

A chronograph measures arrow speed. If you're serious about getting the most from your bow, consider using one to evaluate your tackle as well as your shooting form. Following are some things you can learn from chronograph tuning.

1. Bow Efficiency: Solid plastic arrow rests can produce unnecessary and undesirable friction. Most arrows consistently chronograph two to three feet per second (fps) slower off a solid rest than off a flipper/plunger or similar rest coated with Teflon or shrink tubing. String material also affects arrow speed. On the average, Fast Flight increases arrow speed 5 to 10 fps over Dacron. You can convert most bows from Dacron to Fast Flight. Pro shops have conversion kits.

2. Personal Efficiency: Inconsistent form means missed shots, and you can use a chronograph to analyze and improve your form. For example, creeping or plucking the string, which essentially alters draw length, can cause arrow speed variances of 5 fps or more. Arrow speed should vary no more than plus or minus 1 fps. That is, if base speed is 200 fps, you should be able to hold speeds within the 199- to 201-fps range.

Style of finger protection or release aid can affect arrow speed and consistency. A worn hair tab, for example, not only reduces arrow speed but gives inconsistent speeds, whereas a slick tab improves overall speed and consistency.

3. Kinetic Energy: To gauge an arrow's energy, you must know two variables: arrow weight and speed. Any grain scale, such as a reloader's powder scale, will work for weighing arrows. To compute energy, weigh your arrows, chronograph them, and then apply this formula, in which V equals velocity, W is arrow weight (in grains), and KE is kinetic energy:

$$\frac{V^2 \times W}{450{,}240} = KE$$

With this formula, you can gauge a bow's overall efficiency by comparing KE with total stored energy. Kinetic energy is also a good measure of penetration potential.

Don't climb with your bow.

Tree-Stand Safety

Hunting from a tree can be dangerous if you don't watch what you're doing. Be especially careful in the early morning darkness when you're getting settled on your platform and arranging your equipment. Keep an elbow or some other part of your body in contact with the tree trunk so you don't lose perspective and lean the wrong way. It's a good idea to wear a safety belt, too.

Maximum Range: Bow vs. Rifle

Have you ever wondered how far your gun will really shoot at its maximum range? A 22 Long Rifle bullet will travel approximately one mile—but what about a projectile from your shotgun, handgun or high-powered rifle, or even an arrow from your hunting bow? Comparisons of maximum projectile travel are interesting. Listed below are some of the projectiles more commonly used in hunting, with close approximation of maximum distance each may travel. Keep in mind that an arrow remains lethal at any time in flight (as are most bullets), even nearing terminal velocity. Bullet and arrow weights and resultant velocity will have some influence on maximum travel, as will angle of projectile departure. The following are taken at approximately a 40-degree angle of departure.

Approximate Maximum Projectile Range

BOW OR GUN	YARDS	MILE
Peak bow weight 40 lbs.	240	.14
Peak bow weight 50 lbs.	265	.15
Peak bow weight 60 lbs.	300	.17
12 gauge No. 6 shot	250	.14
12 gauge No. 2 shot	330	.19
12 gauge No. 00 buckshot	748	.42
12-gauge shotgun slug	1340	.76
38 Special pistol	1800	1.0
357 magnum pistol	2350	1.3
22 LR	1760	1.0
222 Rem	3500	1.9
270 Win	4000	2.3
300 Win Mag	4550	2.6

Nature CREDITS
NATURE HAS GIVEN US IN ABUNDANCE —
WILD LIFE
FOOD AND COVER
VAST FORESTS
PRODUCTIVE FIELDS
CLEAN STREAMS
CLEAR LAKES
Mankind DEBITS
MANKIND HAS DESTROYED —
THE VAST FORESTS
POLLUTED THE STREAMS
DRAINED THE LAKES
BURNED THE COVER
AND RUTHLESSLY
DECIMATED WILD LIFE
SPORTSMEN — WHAT WILL YOU DO — HOW WILL YOU HELP AND WHEN WILL YOU COMMENCE TO CORRECT THE ERRORS OF MANKIND AND ASSIST NATURE IN SAVING THE ABUNDANCE OF THINGS ORIGINALLY GIVEN
FEDERAL CARTRIDGE CORPORATION

FEDERAL CARTRIDGE CORPORATION MINNEAPOLIS MINNESOTA

WE MANUFACTURE AND SELL EXCELLENT AMMUNITION

"If you have a son or daughter interested in conservation, send their names. We will furnish a guide to conservation — Fur, Fins and Feathers."

No plan of conservation can succeed without the help of the farmer, his sons and daughters.

Blackpowder

How to Sight In a Muzzleloader

Over normal blackpowder hunting ranges, round and conical bullets have very similar trajectories and can be sighted basically in the same way.

If the center of the group averages about an inch high at 50 yards, the ball or conical will fall about an inch low at 100 yards and about one-half foot low at 125 yards, making 125 yards a practical limit for deer hunting with a muzzleloader. A muzzleloader's rear sight can't be moved up and down, but can be drifted in its notch, making the process very simple:

The Quickloader

Muzzleloader hunters can make handy reloaders for premeasured charges of blackpowder or Pyrodex from discarded plastic cigar cases with snap-off caps. Cut the case to the desired length with a hacksaw, and chamfer the inside of the mouth with a knife to make it easy to reinsert the snap top. A section only two inches long will hold well over 125 grains of powder.

1. Maintain a steady rest from a shooting bench.
2. Shoot a three-shot to five-shot group at 50 yards, trying to print the pattern's center about an inch high of the bull's-eye.
3. To raise the center of impact:
  a. File the front sight down.
  b. Raise the rear sight, if it is an adjustable model.
4. To lower the point of impact:
  a. Buy a taller front sight.
  b. Lower the rear sight, if it is an adjustable model.
5. To move the point of impact to the left:
  a. Drift the front sight to the right.
  b. Drift the rear sight to the left.
6. To move the point of impact to the right:
  a. Drift the front sight to the left.
  b. Drift the rear sight to the right.

Loading to Avoid Misfires

The blackpowder deer hunter loads his smokepole in the morning, but his opportunity may not come until much later in the afternoon. When that chance for a close-range shot presents itself, the last thing a hunter wants to hear is *ffftt!*—the disheartening sound of a misfire. Here is one method of loading that will help prevent that disappointment.

1. Make certain the rifle bore is dry. Run a cleaning patch down-bore with a jag. Run a pipe cleaner through the nipple vent to remove any leftover oil.

2. Ensure that the channel from nipple to breech is clear by placing a cap on the nipple of an unloaded rifle, pointing the muzzle toward a light object such as a nearby leaf on the ground and pulling the trigger to drop the hammer on the cap, firing it. The object should move out of the way. A sharp crack often indicates a blocked vent in the nipple, or other obstruction, while a dull thud denotes a clear channel.

3. Charge the rifle normally, using a powder measure.

4. Before seating the patched ball, place some sort of protection over the powder charge. A preferred material is hornet nesting. Old hornet nests are common. The material may go up in flames to the touch of a match, but in the bore of the rifle, hornet nesting is like asbestos. It will prevent the burnout of the patch and will save the powder from invasion by patch lube.

5. Short-start the lubed patched ball. Generally, a ball close to bore diameter will prove most accurate. Use strong patching material. The short-starter will be required to introduce the tightly fitted patched ball into the muzzle. Then a loading rod or ramrod can be used to fully seat the patched ball on the powder charge in the breech. No need for pounding!

Simply force the patched ball down firmly onto the powder charge.

6. Run a cleaning patch downmuzzle to absorb any leftover patch lube. An oily bore may not shoot to the same point of impact as a dry bore.

7. In rainy weather, carry the rifle upside down to prevent water from seeping into the nipple seat. Also, a device called a Kap Kover can be used to form a gasket over the nipple.

8. Don't cap the rifle until you enter the field. At the close of the day, fire the rifle into a safe backstop, then clean it with solvent and patches, readying it for the next day of deer hunting.

Buck Moves

Early in the deer season, such as during special bow and early muzzleloader hunts, place your stand near feed areas. As the gun seasons open and hunting pressure mounts, hunt close to bedding areas. Bucks eat primarily at night and move away from feed areas by the time it's shooting light.

When gun season opens, keep an eye on bedding areas.

Blackpowder Pour Spout

It's hard to pour blackpowder from the wide mouth of a standard powder canister into a powder measure. A handy pour-spout can be easily made with an empty 38 or 357 cartridge case and the cap from an empty powder canister.

Tap the spent primer out of the 38 case with a nail; drill through the empty primer pocket with a quarter-inch drill; then drill a three-eighths-inch hole through the center of the canister cap. It is then necessary for you to force the 38 case through the hole in the cap from inside the cap until the rim of the empty case jams against the cap. Apply a bead of epoxy glue where the shell rim meets the case, and let it dry before proceeding. Replace cap on canister and pour. When through, replace with closed cap.

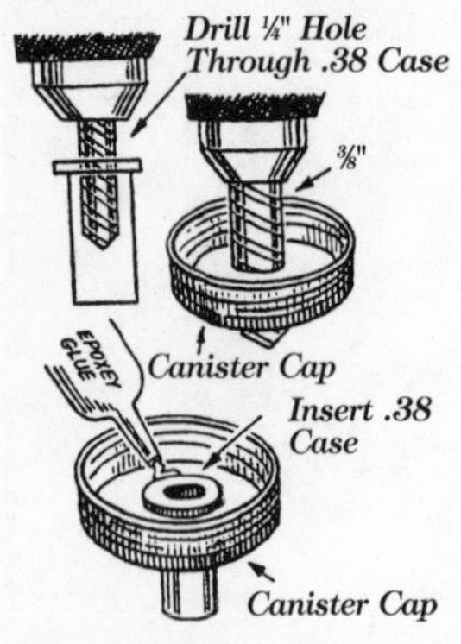

The Finicky Flintlock

Most hunters prefer a percussion (caplock) rifle. However, the flintlock is also fun to shoot, and is the only legal weapon on a few special primitive hunts. Here are steps to aid ignition.

1. Make sure that the rifle is oil-free: Run a cleaning patch down-bore, and a pipe cleaner through the touch hole.

2. Cock the hammer: Lower the frizzen onto the pan. Fire the rifle. A shower of sparks should rain over the pan. Make sure that the flint meets the frizzen face in a flush posture by loosening the flint in the jaws of the hammer (also known as the cock) and then lowering the flint against the face of the frizzen. Then tighten the flint in the cock, with the flint resting flush on the face of the frizzen.

Caplock Rifle

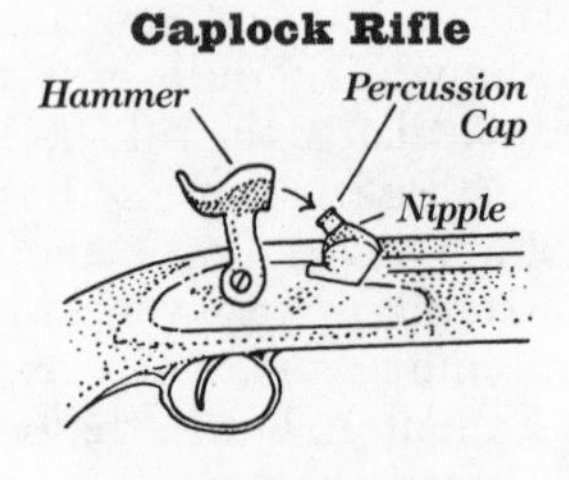

Flintlock Rifle

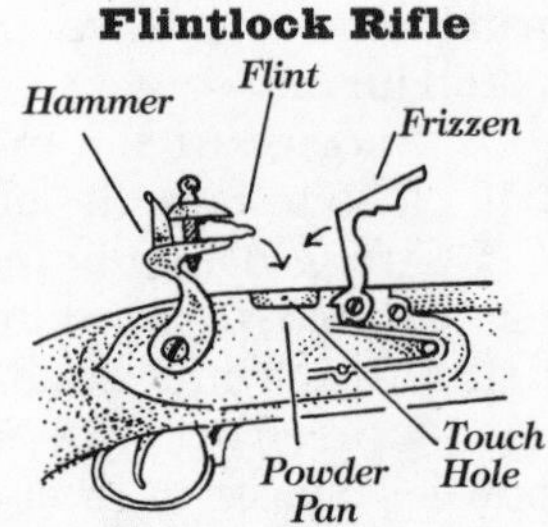

3. Block the touch hole with a pick or a pipe cleaner. Powder in the touchhole has to burn out of the way before the jet of flame from the pan powder can reach the main charge of powder in the breech. Any powder in the touch hole acts as a fuse, slowing ignition.

4. Install the proper powder charge in the breech: cover the powder protector with a very fine material, such as a bit of hornet nesting. Then seat the patched ball firmly.

5. Withdraw the pick or pipe cleaner from the touch hole: Charge the pan with FFFFg blackpowder. Do not fill the pan. Place the powder on the outer portion of the pan, away from the touchhole. One-half to two-thirds of the pan should be filled with FFFFg powder. The idea is to create a flash that travels directly through the touch hole and into the rifle breech. An overly full pan thwarts this cause. If the rifle can be carried slightly tilted to prevent powder from entering the touch hole, all the better. That's not always easy to do in the field, however.

Dry Powder Every Time

Many blackpowder hunts are ruined by a rifle that goes pop when it should have gone bang. Here is a good procedure to ensure that a misfire does not ruin a hunt:

1. Remove all oil from the barrel and chamber. Use a degreaser, or even Coleman fuel, on a patch. A pipe cleaner will remove oil from the nipple area.

2. After the barrel has dried, coat it with wax. Birchwood Casey Gunstock Wax works very well. Make sure the bore is well coated. Remove the nipple and wax the threads, then reinstall it. Wax may also be applied to the entire outer surface.

3. After the wax has dried, load the rifle, being careful not to let any oil or moisture into the barrel.

4. After loading, place a piece of rubber or leather between the hammer and the nipple. Fasten a small plastic bag over the muzzle.

5. During the hunt, carry a piece of plastic, such as

Prune Your Shot

One handy item a hunter should have when in the woods is a small pair of pruning shears. Bow hunters will find a pair invaluable for cutting shooting lanes and building ground blinds. Good pruning shears cost less than $10, fit easily in your pocket, and will cut limbs up to three-quarters of an inch.

Learn how to keep your powder dry, even in snowy or rainy conditions.

a bread wrapper, to place over the hammer area in case of rain.

6. At night, remove the percussion cap, replace the nipple and barrel protection.

7. In a warm tent or cabin, condensation can form in the bore, so consider leaving the rifle outside, perhaps in a vehicle.

There is no need to shoot the rifle empty at night and reload every morning. If in doubt, remove the nipple; if dry grains of powder can be seen in the powder chamber, the rifle will probably fire.

Blaze Orange Nipples

In many states, a muzzleloader is legally unloaded if the percussion cap is off the nipple or the priming powder is removed from the pan. When you unload your muzzleloader in this fashion and leave the charge in the barrel for the next day's hunt, put a scrap of blaze orange surveyor's tape under the hammer. This reminds you not to absentmindedly clear the nipple by firing a cap, which would fire the charge already in the barrel. It also warns you not to load a double charge.

Powder Charges, Grain by Grain

If your muzzleloader is not firing as accurately as you feel it should be, try varying the powder charge you usually load by about five grains more or less. For calibers under 45, FFFg is the most commonly used powder, while most calibers over 44 require FFg.

The greatest accuracy possible with muzzleloading rifles firing round balls can most often be obtained by loading with the maximum possible powder charge. Some maximum charges for round ball rifles are as follows (conical bullets should not be used with these charges):

caliber	grains	charge
32	25	FFFg
34	30	FFFg
36	38	FFFg
38	40	FFFg
40	45	FFFg
42	50	FFFg
44	60	FFFg
46	68	FFg
48	80	FFg
50	90	FFg
52	100	FFg
54	115	FFg
56	125	FFg

F	FF	FFF	FFFF

Aren't Blackpowders All the Same?

Never assume that you can use an old load you find somewhere and get the same results the first shooter got, unless you have the same brand and granulation of powder he had. Different granulations create different pressures in a gun.

For example, it could take 40 percent more FFg to get what you were obtaining with FFFg. But remember that FFFg renders a lot more pressure than FFg.

Never use FFFFg in the breech of a rifle, because it's mainly intended for pan powder in a flintlock, though a few good caplock revolvers have worked with FFFFg. Fg is fine in the 12 gauge, but not much good in the rifle.

FFFg is excellent for the small-bore rifle, most revolvers, some pistols and target loads in big-game rifles. And FFg is good in most big-game rifles with big-game loads. But use the correct granulation. And don't switch brands and expect to get the same ballistic results you were getting with the other powder.

Sprue Up or Sprue Down?

Some hunters spend countless hours searching out and debating the factors of muzzleloading accuracy (patch thickness, patch material, patch lubricant, variable powder charges, etc.). Very little time seems to be devoted to one of the most obvious variables. Most hunters take a pure lead ball, set it sprue up on the greased linen patch, and push it firmly home on the powder. Why sprue up? Because

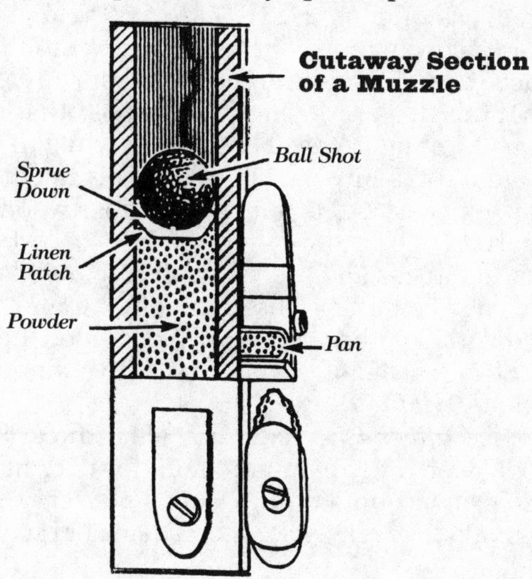

common knowledge suggests you can use the sprue as an index mark to ensure uniform bullet position from shot to shot. Is it true? Probably not.

When a ball is loaded with the sprue down and tested carefully from a benchrest, accuracy is almost always better than when the bullet is loaded sprue up. This is particularly true at ranges in excess of 50 yards, and the reason is obvious. When loaded sprue down, the round ball is, of course, aerodynamically superior to the ball that is loaded sprue up. Simply put, it is more streamlined.

If the shooter needs a method to index the ball, this can be accomplished by making a tiny punch mark or file nick at the bottom of the bullet mold, directly opposite the sprue location. When loading, place the ball with this tiny mark uppermost. For added consistency, align the mold seam mark with the axis of your front sight blade. Wait for a calm day, and then shoot five or 10 shots from a rest, at a 100-yard target. You will be in for a pleasant surprise.

Blackpowder Paraphernalia

Success at muzzleloader deer hunting demands a basic knowledge of one's quarry and proven hunting techniques, as well as the various types of equipment. Blackpowder hunting is a game steeped in gear. Here's an in-depth look at the paraphernalia you'll need to be successful.

✻ ***Rifle Choice:*** You can take any number of routes in choosing a hunting smokepole—from authentic replicas of guns used 100 to 200 years ago to ultra-modern versions. The replicas are more historically accurate, but most hunters opt for a cleaner, more subtle, modern-looking smokepole that meets serious hunting needs. They'll choose a gun with a minimum of shiny metallic embellishments that can spook game and a simple, low-gloss wood stock. The barrels will be short—in the 24- to 28-inch range—for lightness and easy maneuverability in thick cover, where the biggest bucks are often found. And they'll go with a percussion, or caplock, model (invented in 1807) instead of the more primitive flintlock (around since 1615).

✻ ***Caliber Choice:*** A 45-caliber firearm should be considered the bare minimum for deer hunting. This caliber accounts for thousands of deer yearly, but you should take only perfect broadside or quartering-away

shots with it, and not stretch your range. A better choice by far is the 50, or even the 54 or 58.

✳ ***Barrel Twist:*** Rate of rifling is another factor to consider when choosing your muzzleloader. This provides the projectile with a stabilizing spin that helps ensure downrange accuracy. The rate of twist you choose depends on whether you'll be shooting roundballs or conical bullets. Roundballs perform best with a slow rate of twist—say, 1:66 (one turn in 66 inches). Conical bullets shoot better when the barrel twist is more severe, such as 1:34. For a good compromise twist for both, go with 1:48.

Help From Hornets

Muzzleloader shooters who prefer to shoot patched round balls sometimes have difficulty finding patching material that fits their rifles perfectly. A poor fit can allow hot powder gases to blow by the patched ball, ruining accuracy. Hornet's nest material can solve this problem. Tamp a few layers of the nest material on top of the powder charge, then seat the patched ball. The patch will be saved from powder-gas burnout, and accuracy will improve. Hornet's nests are best collected in the fall, after the hornets have abandoned them.

✳ ***Working Up a Load:*** If you have a 50-caliber rifle, shoot a few rounds with 50 grains of powder, then slowly work up, never exceeding the recommended maximum. At some point your accuracy will fall off. You want to pick a load below that level.

✳ ***Tools of Muzzleloading:*** Besides gun, bullets and powder, you'll need several other items to hunt deer with a muzzleloader. Bullet lubricants are necessary, unless you buy prelubricated versions. Patches are also required for roundballs. You'll need a short (ball) starter to poke the bullet into the barrel when you first load up, and a ramrod to drive the projectile down until it lies snug against the powder charge.

Caps are required for ignition, and these come in many sizes, so be sure you choose ones that fit over the nipple of your gun. Most muzzleloaders take a No. 11 cap. If you're shooting a flintlock, you'll need spare flints and FFFFg priming powder.

A graduated cylinder allows you to accurately prepare powder charges, but most sportsmen like to prepare several complete loads and have them ready for use in the field. Premeasure the amount of powder you need and also have a lubricated bullet and cap ready for fast reloading.

You'll want to have a spare nipple, combination nipple wrench-pick, and a ball-puller in case you ram a bullet down your barrel without first putting in a powder charge and need to get the projectile out.

✳ ***Cleaning Up:*** After you shoot a blackpowder rifle, you *must* clean it. If you don't, the powder will cause rust and corrosion, ruining the firearm. Use hot, soapy water, scrub all metal parts, and scour the barrel with swabs until the water comes out completely clean. After the gun is dry, apply a light coat of oil to the inside of the barrel and exposed metal parts.

STEVENS

THE difference between Hitting and Missing is the difference between an Accurate and an Inaccurate ARM. Choose wisely—discriminate! Get a STEVENS! Forty-three years of experience is behind our *tried* and *proved* line of

RIFLES, PISTOLS, SHOTGUNS, Rifle Telescopes, Etc.

Ask your dealer and insist on the STEVENS. If you cannot obtain, we ship direct, *express prepaid*, upon receipt of Catalog price.

Send 4c. in stamps for 140-page catalogue, describing the entire STEVENS line. Profusely illustrated, and contains points on Shooting, Ammunition, Etc.

Beautiful Hanger will be forwarded for 6 cents in stamps.

J. STEVENS ARMS & TOOL COMPANY,
P. O. BOX 5680,
CHICOPEE FALLS, MASS., U. S. A.

6

Hunting Safety and Survival

So if you must nap while nimroding, Brother Hunter, don't do it beside a dead deer; or a log or stone that looks like a deer; or on a used deer trail; or just don't do it. Yet if mayhap you must, pick a spot well out in as open a place as you can find, make yourself quite visible from all angles, and run up a red flag or two. But I still can't rate you 100% if you snore.

—S. Omar Baker,
***Sports Afield Hunting Annual*, 1935.**

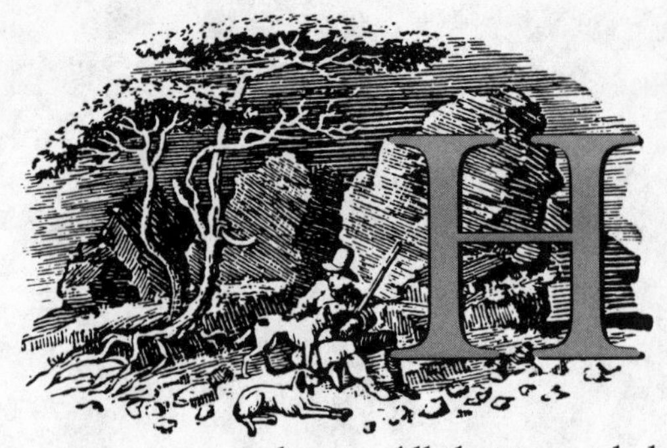

How Cold Can Kill

Hypothermia, a lowering of the core temperature of the body, kills deer hunters and other outdoorsmen each year. It can strike in any season, in almost any climate. All that is needed is air temperatures of 30°F to 50°F, wetness (rain, sweat or a dunking), a slight wind and a tired hunter.

Hypothermia attacks in two stages. First, your body begins to lose heat faster than it can produce it. At this point you are aware of feeling cold, and shivering begins.

The second stage is when the cold reaches the brain, depriving you of judgment and reasoning power. Because of this, almost no one recognizes that he has hypothermia. In this second stage, your internal temperature is sliding downward. Without treatment, this slide leads to stupor, collapse and perhaps death.

There are several ways to avoid hypothermia:

✳ ***Stay dry.*** When clothes get wet, they may lose as much as 90 percent of their insulating value. Wool and some high-tech insulators, however, will retain most of their warmth; cotton, down and some synthetics lose more.

✳ ***Beware of the wind.*** Even a slight breeze carries heat away from bare skin much faster than still air. Wind drives cold air under and through clothing, refrigerating wet clothes by evaporating moisture from the surface.

✳ ***Understand cold.*** Most outdoorsmen simply can't believe that air temperatures of only 30°F to 50°F can be dangerous. Some fatally underestimate the danger of being wet at such temperatures.

✳ ***Terminate exposure.*** When you can't stay dry and warm under existing weather conditions, be smart enough to call it quits and return to camp or home.

✳ ***Never ignore shivering.*** Persistent or violent shivering is clear warning that you are on the verge of hypothermia.

✳ ***Learn to recognize the symptoms.*** Uncontrollable fits of shivering, chattering teeth, etc.; vague, slow, slurred speech; lapses in memory; immobile or fumbling hands; staggering and stumbling; drowsiness; exhaustion; inability to get up after a short rest.

In most cases, the victim will deny he's in trouble, but believe the symptoms, not the victim. Treatment should be immediate.

First, get the victim out of the weather and remove his wet clothes. If he is only mildly impaired, give him warm drinks and get him into dry clothes and a warm car, room or sleeping bag. If the victim is semiconscious or worse, he does not have the capability of regaining his body temperature without your help. Keep him awake, give him warm drinks, and undress him. Put him in a sleeping bag with another person, also stripped. Skin-to-skin contact is the most effective treatment in this desperate situation.

Hypothermia has been called the killer of the unprepared. So on your next trip, go prepared.

Frost Nip and Frost Bite

Most cases of frost nip are nothing more than a burning sensation at the tip of the nose or the edge of the ear. Usually you can relieve the pain by touching it right away and shielding it against the cold.

Serious frostbite is often due to wet socks or gloves. After the first painful coldness, there is a progressive loss of sensation, and you may not notice anything wrong again until stiffness and loss of function set in.

Serious frostbite should be treated by rapid rewarming in a tub or sink. The water temperature should be kept, if possible, between 38°C and 40°C (100°F to 105°F). Since the victim can be easily burned, be sure it is not too hot. Pain pills (such as two aspirin) should be given even if the victim has not yet complained, because the pain will come on sharply and suddenly.

Continue rewarming until the skin looks pink (redder than normal) and the tissues feel soft—about 30 to 45 minutes. Do not rub the affected area. After rewarming, gently dab the area dry, cover it with sterile dressings, and keep it at rest, with the fingers or toes separated with cotton or gauze.

Skin blisters usually develop in a day or two and

The Disoriented Compass

Hunters who rely on a magnetic compass for direction are well advised to hold their guns at arm's length before taking a reading. The amount of iron in the average firearm is significant enough to alter the true reading of a compass by several degrees. Likewise, other articles that contain iron (knives, some belt buckles and watchbands) can also throw your compass off and consequently give you a false reading.

are not necessarily a bad sign. If the blister fluid becomes cloudy or bloody, however, it is probably an infection and should be treated with prescription antibiotics.

What Do You Know About Ice?

Thick ice is not necessarily safe, and dark-colored ice isn't necessarily dangerous. Knowing the basic types of ice and their major characteristics could save your life. See if you can match the ice types with their characteristics at right.

A. Pack ice	***1. Appears milky; is porous and generally dangerous.***
B. Candle ice	***2. Striped appearance caused by layers of frozen and refrozen snow. Only as strong as underlying ice.***
C. Clear ice	***3. Wind-blown into piles. Hollow areas or holes often exist.***
D. Snow ice	***4. Usually appears in late winter or spring. Many vertical lines bundled together, indicating deteriorated and weak condition.***
E. Frazil ice	***5. Blue, green or black. Often the result of a long, hard freeze. With four or more inches of this type, foot travel is generally safe.***
F. Layered ice	***6. First ice. Surface may appear oily or opaque and is composed of disk-shaped crystals. Extremely dangerous.***
G. Polyna ice	***7. A nonlinear opening in the ice. Annually persistent on a given lake or river area. An obvious hazard.***

ANSWERS: A.3, B. 4, C. 5, D. 1, E. 6, F. 2, G. 7.

Indian Snow Shelter

Less familiar than the igloo—but far more practical in the deep, unpacked snow typical of many forested regions of North America—is the quin-zhee, a shelter traditionally built by Athapaskan Indians. It is constructed by heaping loose snow in a cone-shaped pile five or six feet high, allowing it to settle for an hour or two, then hollowing out the inside.

The hour or two of undisturbed settling time is

important, because it allows the snow to undergo a process known as destructive metamorphism, during which snow crystals bond, creating naturally packed snow that makes it possible for the structure to be self-supporting.

A person working alone and equipped only with a snowshoe for a shovel can complete a quin-zhee in a couple of hours, building a strong, well-insulated emergency shelter that quickly becomes as much as 40°F warmer than the outside air.

A Little Light on the Subject

A good number of kills come just as the sunlight begins to fade. Low-light or no-light conditions make field-dressing extremely difficult, so the wise deer hunter should always carry his own light source. Lightweight, adjustable-beam flashlights can be easily stowed in a jacket pocket; better yet is the type that clips to the bill of a hunting cap. The hat clip also functions as a base so the flashlight can be positioned on the ground, a stump, rock or other surface, and the light directed just where you need it.

Safe Snowshoeing

Outdoorsmen are snowshoeing again. Many sporting goods stores all over the country are getting requests for the rawhide shoes. Why? For one thing, you don't need lessons—anyone can learn to walk with snowshoes in minutes. For another, snowshoeing is safe. If you fall, the harness (or binding) lets your foot go at once.

Snowshoe equipment is inexpensive and in many places you can rent a good pair of snowshoes with harness. The charge for the weekend is usually less than $10. At Army-Navy surplus stores a pair of snowshoes cost anywhere from $25 to $40. A coat of shellac or varnish once a year will keep them in good repair for a lifetime. Also the webbing can be tightened and/or replaced with little trouble.

Here are tips to keep you from getting caught in unsafe situations:

✱ Don't travel during a blizzard. Always check the weather forecast before starting.

✱ Clear winter air makes estimating distance tough. Remember that underestimates are more frequent than overestimates.

✱ Avoid traveling in whiteout conditions, when lack of contrast makes it impossible to judge the nature of the terrain.

✱ If you camp overnight, stop early enough to have plenty of time to build a shelter.

Sure Signs of Snow

To our forefathers, knowledge of impending snowstorms and hard winters meant much more than buying extra groceries. For them it was often a matter of survival.

The family hearth was a prime place to ascertain coming snow. A wood fire that hissed and made sounds similar to those of someone tramping throught heavy snow meant a storm was on the way. So did whitened ashes moving, seemingly of their own accord, around the fireplace. Smoke that hugged the ground indicated deteriorating weather conditions.

Other snow signs were connected with the behavior of domesticated and wild animals: cats sitting with their backs hunched up to the fire, cows coming to the barn in the middle of the day, deer seeking out thickets, and "bunny barometers"—cottontails uncharacteristically abandoning their beds during the middle of the day and moving to brushpiles or groundhog holes.

Fowl behaving strangely suggested that snow was imminent, as did birds fluffing out their feathers and feeding ravenously, snowbirds gathering in bunches on the ground, roosters crowing at midnight, a whitish ring around the moon on a frosty evening, clouds gathering after a frosty morning, or herringboned layers of clouds and a purple sky on the evening horizon. For long-range forecasting, the number of morning fogs in August determined how many big snows a winter would have, while thick corn shucks, heavy layers of fat on bears, heavy mast crops and extra-dense coats on furbearers all heralded a hard winter.

Watch That Muzzle

You've seen photographs of a hunter holding a rifle with one hand around the barrel a couple of inches back from the muzzle. The receiver is resting on his shoulder at a point just ahead of the triggerguard, and the butt hangs out behind his back. This appears to be a safe way to hold a rifle, but if he should slip and fall backward, his reflex will be to extend his arms out and back. The muzzle ends up pointing into his ribs. If the gun discharges, the resulting wound will likely be fatal.

First-Aid Essentials

Simple mishaps in the outdoors can usually be handled with a homemade first-aid kit. A tacklebox or fanny pack makes an ideal container. The Louisiana Department of Wildlife and Fisheries recommends you carry these basic items in your kit: allergy or prescribed medication, antacid, antihistamines, antiseptic wipes, assorted adhesive bandages, bandannas, bar soap, burn ointment, butterfly bandages, cotton balls, disposable latex gloves, elastic bandages, eyewash, first-aid cream, first-aid guide or booklet, gauze bandages, hydrogen peroxide, insect repellent, instant hot/cold pack, itch ointment (calamine lotion), medical tape, moleskin, painkiller (aspirin), scissors, snakebite kit, sunscreen cream, tongue depressors and tweezers. Add other items according to the type of activity you'll be engaging in.

Remember to update your kit by replacing missing or used items, and by checking the expiration dates on medicines. If you haven't taken a first-aid class, enroll in one.

Trick Wicks for Survival

Trick birthday-cake candles—the kind that won't go out no matter how hard you blow on them—make excellent windy-day fire starters for camping. Just nestle one under some kindling and light. Wind gusts won't blow it out, and during the calm periods the fire will be set. You'll find the candles in novelty/joke shops.

Emergency Gear

Accidents and emergencies will happen, sometimes before you even get into the woods. For sportsmen who do a lot of wintertime traveling, here are a few things that you might want to have in the car, just in case.

✳ ***Jack and Lug Wrench:*** It's a good idea to practice with your jack at home to see how it works.

✳ ***Tire Chains:*** A must for wintertime travel.

✳ ***Tire Pump:*** There are all kinds, from hand pumps to foot pumps. You can even buy a pump that can be plugged into your cigarette lighter.

✳ ***Flashlight:*** You can also buy a special light with a long cord that plugs into the cigarette lighter. Flares are helpful in highway emergencies.

✳ ***Fire Extinguisher:*** Buy the all-purpose type and check the pressure indicator on it from time to time.

✳ ***Booster Cables:*** Check your car manual on how to hook them up. Failure to hook cables properly can result in an exploded battery.

✳ ***Hand Tools:*** Open-end wrenches, pliers and screwdrivers should get you by, but you may want to add to that assortment. A shovel comes in very handy at times.

✳ ***Gas Can:*** Some motorists also carry a siphon with a squeeze bulb that enables you to get fuel from another motorist.

The Ultimate Survival Kit

✳ ***Space blanket:*** one side blue, the other side reflective silver. ***Long underwear:*** upper and lower. ***Heavy wool socks. Heavy wool cap with face covering. Thermal gloves. Full rainsuit with hood. Down jacket. Booties and overalls. Swiss army knife.***

Signal mirror. Signal whistle. Roll of toilet paper. Insect repellent. 3 small cigarette lighters. Strike-anywhere wooden matches: cut in half and stored in a waterproof container. ***3 paraffin candles. Compass and maps of the area. Pencil and paper. Heavyweight plastic garbage bag. 100 feet of nylon cord. First-aid kit. Water purification tablets. One-quart plastic bottle:*** hint—put your first-aid kit in the bottle. ***Granola bars:*** at least one dozen. ***Tube tent:*** fluorescent orange, with stakes. ***Heavy-duty aluminum foil:*** 10 feet. ***Navigation tape:*** 2 rolls, fluorescent orange. ***Fishing gear:*** one dozen snelled Size 14 hooks, 5 Mepps 00 gold spinners, 20 split-shot sinkers, jar of salmon eggs, 5 artificial flies, 100 yards of 6-pound monofilament. ***Monofilament:*** heavy, 40-pound-test or more, for snares and small-game nooses. ***Cook kit:*** aluminum cup, plate, pot with lid, and 8 x 8-inch cooking grate.

Eyeglasses for Hunters

According to the American Optometric Association, regular eyeglasses may do hunters more harm than good because the prescription lens may not be set high enough for sighting a gun. Also, glasses with small lenses or thick frames and wide temple pieces can hamper side vision.

Hunters should tell optometrists about their sports and obtain frames and lenses specially suited for hunting. Hunters with poor eyesight, for instance, should use telescopic sights. With a regular sight, shooters must focus from back to front sight to the target in a fraction of a second. Hunters with weak eyes may find it difficult to focus on the back sight, and a telescopic sight helps by providing a close, clear look at the target.

Hunters should also protect their eyes from powder burns, spent shell casings, branches, flying shot and other debris by wearing safety goggles or prescription safety eyeglasses.

Self-Help for Heart Attack Victims

A heart attack victim who collapses in public will often be saved by a bystander who knows how to give CPR (cardiopulmonary resuscitation). But what happens to the heart attack victim who is all alone? Doctors say he probably has only about 10 seconds before he loses consciousness.

According to a book titled *Emergency Medicine*, heart attack victims can help themselves by repeated coughing. A very deep breath must be taken before each cough, and the cough must be strong and prolonged. The breath and cough must be repeated every one to two seconds without a break until help arrives or until the heart beats normally again.

Deep breaths bring oxygen into the lungs, and the coughing movements squeeze the heart, keeping the blood circulating. A heart attack victim can get to a telephone and, between breaths, call for help.

Pulmonary Edema

Sportsmen should know about a serious altitude-related illness called pulmonary edema, which can occur above 9000 feet. The lungs fill with liquid, which leads to a lower-than-normal concentration of oxygen in the blood. If ignored, it can result in death.

According to Dr. John Reeves, a researcher into altitude-related disorders at the University of Colorado Health Science Center in Denver, initial symptoms include shortness of breath and an irritating cough. A bluish discoloration of the lips and fingernails often follows. Undue fatigue, coughing up of a white, frothy sputum, headache and nausea may appear as the condition worsens. Eventually a loss of mental acuity occurs. If the victim lapses into unconsciousness, death may follow in as little as two hours.

Researchers aren't certain of the mechanisms at work, but your chances of suffering this malady depend on several risk factors, including speed of ascent—rapid, strenuous ascents are dangerous; the degree of cold—extreme cold weather increases risk; and exertion—heavy work at high altitude increases risk. A cold with chest congestion will make all of the above worse.

There are no hard-and-fast rules for prevention. Still, proper acclimatization appears to help. Physical fitness has nothing to do with the slow, cumulative process necessary for acclimatization. Gaining 1000 feet per day is ideal.

Low-Rent Tent

An efficient, inexpensive shelter can be made from a 9 x 12-foot tarp, available in most hardware stores for $10 to $20. A waterproof plastic coating and brass grommets make them ideal for the occasional camper. To set it up, tie a crosspiece horizontally between two trees about five feet apart. Prop three poles or straight, sturdy branches against the crosspiece so they angle to the ground for support. You can use stones and logs to anchor the sides of the tarp to the ground.

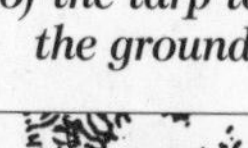

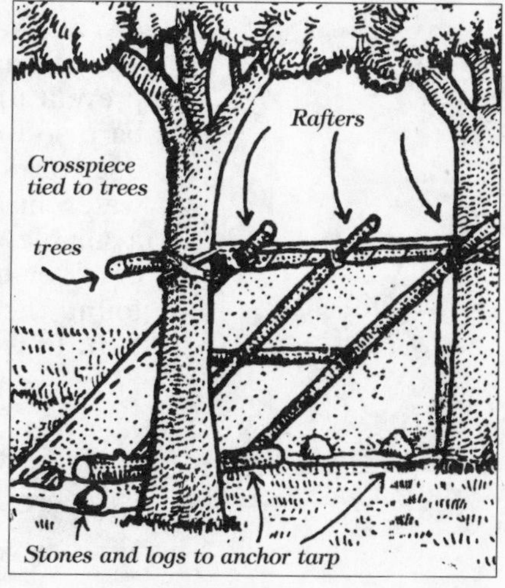

A Hunter's Hat

When the leaves are falling or the snow is flying, don't put away those long-brimmed baseball caps.

Under the diffused light conditions of an overcast day, your pupils tend to constrict. This constriction reduces your ability to define color and shadow. A baseball cap eliminates skylight, allowing your pupils to dilate, and in effect increases eye efficiency, letting you spot game more easily.

If a deer stand requires you to face south into a low winter sun, the brim will also prevent glare-induced headaches. And for those who wear eyeglasses, the cap acts as a deflector, protecting the lenses from rain and snow.

Out on snow-covered ice, reflected and direct sunlight are nearly equal. While sunglasses offer significant relief, their effect is nearly doubled with the addition of a baseball cap.

When it is bitterly cold, pull a stocking hat over your cap. While this combination might not appear fashionable, it will effectively hold in your body heat.

Burn Classifications

First-aid kits usually provide directions for the treatment of first-, second- and third-degree burns. But do you know what each is? Here are the official classifications:

First: *The outer skin is reddened and slightly swollen.*

Second: *The underskin is affected, and blisters form.*

Third: *The skin is destroyed and tissue underneath is damaged.*

Learn these classifications so you can identify each and administer the proper treatment.

Seven Secrets of a Smart Campfire

1. We all know that hardwoods make better fires than softwoods—but which is which? The heavier the wood is when dry, the better it is for a campfire. Even softwood will burn satisfactorily if it hasn't started to rot and is thoroughly dried.

2. If a fire ring is available, use it. If not, choose a spot away from any trees and scrub, and keep well away from dead trees, stumps and exposed roots to prevent underground fires. Build a fire ring only on bare soil or on top of a rock or gravel base. Never use rocks gathered close to water because trapped water may make them explode. If dry rocks are available, build a large fire ring so the rocks won't crack or blacken and can be returned to where you found them.

3. In wet weather, build the fire on a slight mound of earth or on a flat rock. This will provide drainage in case of rain.

4. In dry weather, or if rocks aren't available, dig a shallow fire pit and use the earth removed to build a low windscreen around the pit. Before starting a fire, soak the perimeter of the pit with water.

5. In any weather, build a small fire. It'll cook food better, conserve wood, and minimize the danger of a forest fire.

6. Each time you make a campfire, dry enough wood for the next one. After it cools, stash it under a plastic cover or inside your tent.

7. Always keep a large pot or some container of water close at hand for emergency use. And when you're ready to break camp, drown the fire. Then drown it again. You'll never be sorry you did.

Making Fire Starters

If you spend a lot of time in the woods, it's prudent to make a large supply of fire starters and keep them on hand.

My favorite fire starter is made from paraffin blocks. Over a low flame melt the paraffin in a two-pound coffee can, with the top bent to form a spout. Other waxy materials such as candle stubs or bits of crayons can be recycled into the mixture.

When the brew is ready, dip pieces of string in it and lay them out on a newspaper. They stiffen immediately, and you can then clip them into two-inch lengths for wicks. Pour more melted wax into two- or three-ounce paper cups, filling them about three-fourths full. As soon as the wax starts to congeal on the surface, insert the wicks. After the fire starters have cooled and hardened, store them in plastic bags. Three or four of these fire starters, along with a butane lighter, will always come in handy.

Fire Starter

One of the cheapest, most efficient fire starters you can take camping is a 2-foot-long 2x4. With nothing more than a good pocketknife, you can easily whittle a pile of dry kindling in no time. Stuff this beneath a few twigs and limbs, and you should be able to start a fire quickly.

A Heating Hole

With this handy trick, used by frontiersmen in the East, you can make that long wait for game in bad weather as comfortable as waiting in a warm room. No doubt passed down to them by Indians, it was very popular among hunters who provided meat for forts under siege. These men had to stay both warm and unseen, as the line between hunter and hunted in those days was thin.

Slipping out of the fort in the dead of night, hunters would go to preselected spots to wait for game. If it was extremely cold, they found a white oak tree and peeled off the bark in slender pieces. They then dug a hole in the ground, roughly nine inches deep and six inches wide, and filled it almost to the top with crisscrossed layers of bark. Covering themselves and the hole with a robe or blanket so as not to attract attention, they started the bark smoldering, using flint and steel to spark the fire. Once the bark was burning well, they covered the pit with earth, leaving two small air holes.

In this fashion they could sit cross-legged, with the heating hole between their legs. As white oak burns with little or no smoke, the hunters spent the rest of the night and early-morning hours in comfort—some even slept—with little worry of being discovered by hostile forces or of freezing to death.

Abundant Tinder

Much has been said about finding sources of dry wood when camping, and any boy who has knotted a neckerchief knows enough to always bring along a match case or a lighter, but all outdoorsmen should know and be able to recognize the following types of natural tinder:

✳ ***Outer Bark:*** Birch and cherry trees are a great source of combustible bark that burns with a persistent, sputtering flame. Gather the bark that peels naturally (like skin after a sunburn) from these trees' trunks. Warning: Do not strip the bark from live trees. Not only will you destroy the tree, but you'll find live bark to be wet and useless tinder, anyway.

✳ ***Fungi:*** Several types of fungi have a fluffy layer beneath the cuticle that was at one time collected and sold as "German" tinder. Dried fungi burns very well, and it's wise to keep an eye open for it when hiking in to your campsite. Though it's difficult to find in any real quantity, a handful will do if you're in a jam.

✳ ***Saps:*** Depending on the part of the country where you camp, you'll find that cone-bearing trees such as pines and firs exude sap from their trunks and branches. In a pinch, a sticky evergreen branch can be used to kindle a fire. These branches tend to burn hot and quickly, so you rarely need much to get the job done.

The time to collect tinder is not at dusk, when you've pitched your tent and are eager to get the trout frying. Collect during the day, as you hike, fish, or hunt. Store whatever natural sources of tinder you

Dry Firewood

In rainy or snowy weather, it's frustrating trying to keep firewood dry by covering it with a tarpaulin or sheet of plastic that the wind soon blows away. Instead, use the "rooftop" stacking method, in which the rounded, bark-covered splits of wood are all on top. In this manner, the stacked firewood will shed even the most torrential downpour, just like shingles on a roof, thus keeping the inner heartwood almost completely dry.

find in a large Ziploc bag that can be sealed, folded and tossed into a daypack or jacket pocket.

Foods for the Woodsman

Anyone considering a serious backpack hunting trip should research food, mineral supplements and vitamins. You'll need foods that are high in complex carbohydrates, and are already prepared or easy to prepare. Good examples include pastas, whole wheat and bran muffins, various grain cereals, dried fruits or fruit roll-ups, dried potatoes and similar foods rich in starch. When you are exercising heavily, you need to eat carbohydrates to quickly replace that used-up energy. Protein-rich foods are good for building and repairing body cells, but they do not supply quick energy. The body uses existing body fats and carbohydrates for quick energy needs.

In selecting trail food, shy away from the common nuts and jerky. These items are hard to digest and do not provide the quick energy you need. Instead, take such items as raisins, dried oats, figs, sunflower seeds, dried banana chips and shredded coconut.

It's important to drink lots of liquids, and if you have space, packaged juices such as spiced cider are tasty. A little caffeine is also good for the system.

✳ ***Remember what the cross-country skiers and long-distance runners have learned:*** It's not what you eat on race day that is the most important; it's what you eat the night before. It's also important to do no strenuous exercise for at least a week prior to your hunt. This lets your body store up extra fat reserves.

✳ ***Don't forget your vitamins.*** Large doses (1000 milligrams a day or more) of vitamin C and dolomite, taken for at least a week before and all during the hunt, will help fight sickness, help the body to repair torn muscle fiber caused by strenuous hiking, and help combat the muscle stiffness and soreness associated with hunting. Your friends may laugh at you when you first start this procedure, but it won't be long before they'll be asking you for pills.

Warm Your Ax

Cutting firewood in freezing weather is one sure way to warm yourself up. But your ax should be warmed up beforehand by putting the blade near the campfire. Steel can become brittle in subfreezing weather, and the ax just might be ruined if you hit a knot. Do not put the blade directly into the flames; if you do, you'll ruin the temper of the steel.

Hot Toddies Are Not So Hot

Winter sportsmen looking for the perfect cold-weather antifreeze are not likely to find it at the liquor store. Alcoholic beverages act as vasodilators, opening the blood vessels and causing a surge of blood to the capillaries near the surface of

the skin. That surge causes the sensation of warmth once thought to be beneficial. In fact, there is a big price to be paid for that moment of warmth. The heat released to the skin is coming from the core of the body, so the overall body temperature declines.

In cases of extreme temperature loss, alcoholic beverages can be life-threatening. Dangerously chilled people may slip into comas or die when given liquor.

The best cold-weather beverages are still hot liquids. A cup of steaming soup or tea will go a lot farther toward making that 10-below-zero day bearable than will a snort of Old Gun Shot.

Tricks With a Tarp

A tarpaulin—a rectangle or square of waterproof material equipped with eyelets on all sides—provides the simplest form of shelter for outdoorsmen. Here are four you may find useful:

* ***Tepee:*** Rig four poles to a peak so that lines drawn from base to base make a square. Wrap your tarp around it, and tie it down with thongs to the eyelets.

* ***Lean-To:*** Cut two crotched poles six feet high and two that are three feet high, as well as two long poles to lay across both sets—six poles in all. Set up both sets parallel, six feet apart. Run your tarp from the high ridge back and down over the low ridge to the ground. Guy all four crotched corners so they will pull the tarp tight.

* ***Forester's Tent:*** Cut a pair of sheer poles, tie them together at the top, and spread them so the crotch will be about six feet above the ground. Run a long pole from the point where it rests in the crotch down to the ground at the rear. Sling your tarp over the structure, and weight, stake, or tie down the sides,

folding them under as you near the back end of the tent.

***A-Tent:** Finally, of course, there is the open-ended A-tent, which is so simple it requires no explanation.

Space-Age Warmth

A valuable piece of equipment that has been on the market for some time is a space blanket. Mine looks like a 7 x 4-foot piece of aluminum foil and weighs about two ounces. It can easily be folded and tucked into a hunting jacket pocket.

In cold weather the space blanket can be placed on top of an air mattress or ground cushion to reflect the body's heat. I have used mine in 20°F temperatures on top of an inflated rubber air mattress, and it felt like an electric blanket. It should not be used on top of your sleeping bag at very low temperature, however, because moisture will condense under the blanket and saturate the bag.

In summer, when sweating isn't a critical problem, the foil can be used as a lightweight emergency blanket or rain shelter. Once while I was camping, a thunderstorm flattened my tent, but I stayed warm and relatively dry by covering up with my space blanket.

When Lightning Strikes

Lightning isn't all that bad. Even if you get hit with 50 million volts, it is direct current, which is a lot better for the heart than alternating current.

Alternating current causes the heart to fibrillate, or quiver, which is hard to stop. The direct current from a lightning bolt will make the heart stop, but then the normal beat can resume, just as it does when a person is shocked during a resuscitation.

The two likely causes of death from lightning are direct brain damage and respiratory paralysis. Little can be done for brain damage, but simple artificial respiration can save the life of someone with respiratory paralysis. The heart will resume beating; if it doesn't, do CPR—cardiopulmonary resuscitation.

The rule of thumb for resuscitating a person who has been struck by lightning is to perform CPR until help arrives. If there is a heartbeat, mouth-to-mouth resuscitation is all that is needed. The second rule is to treat even those who appear dead, since there is always a chance you can save them. A person who shows any signs of life after having been struck will almost certainly survive.

Preventive Moleskin

Maybe the load is heavier than usual, maybe the terrain steeper, or possibly the boots have just seen better days. At any rate, you get a blister. Moleskin can help ease the pain and protect hot spots so you can finish the day. But why not get in the habit of putting the moleskin on before you start to walk? An application to the heels of both feet and any known problem toes should keep you from having to stop. Especially for sportsmen who get out infrequently, this treatment can save a lot of pain and grief.

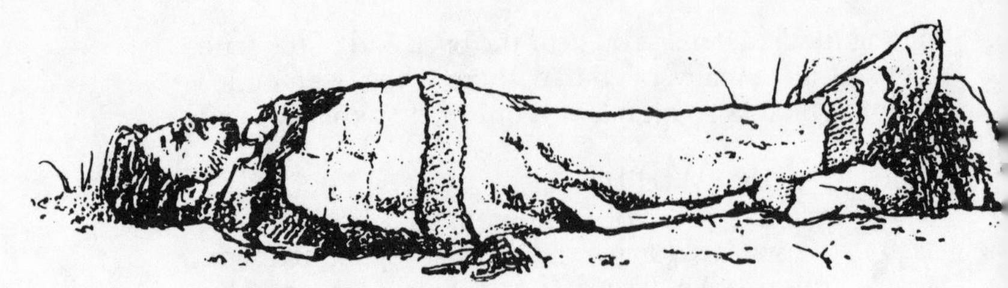

What You Must Know to Survive

Wilderness survival skills can be broken down into five areas. Every outdoors-oriented person should be skilled in each.

1. First Aid: Studies show that a large percentage of survivors had to perform first aid on a fellow survivor, or themselves. Even a trivial wound or illness can become life-threatening. If you haven't done so, take a course through the American Red Cross.

2. Signals: Always carry a selection of signal devices with you, both audio and visual. A sturdy plastic whistle is an example of an audio signal, as is your firearm. A shatter-resistant signal mirror and a compact pop-up flare are examples of visual signals.

3. Fire: Fire can warm you, cook your dinner and boil water, signal a would-be rescuer, instill a feeling of well-being and perform many other services. Always carry a waterproof container of strike-anywhere wooden kitchen matches and other fire-starting devices, plus a 35mm film canister stuffed with cotton balls, which make great tinder.

4. Food and Water: Long-term survival requires that you find food. A healthy knowledge of edible plants helps, as will your ability to set effective traps and snares for small game. Water, generally speaking, is more important than food when it comes to survival. Carry some extra, and be alert for signs of nearby water, such as swarming insects or birds heading in one direction, especially at dawn and dusk. Look for convergent game trails, and remember depressions and draws depicted on your topographic map.

5. Shelter: Building a snug shelter can mean the difference between a miserable night afield and an enjoyable one. But before you set to work on a lean-to or three-pole A-frame shelter, look for one that nature has already prepared for you, such as a rock overhang or a hollow log.

Winter Shades

Blue sky, white snow and bright sunshine can produce a beautiful winter scene, but they also form a combination that is hard on your eyes and potentially dangerous. Most people know that continued exposure under these conditions can cause snow blindness, but what they don't consider is night blindness. The glare from the sun and snow can reduce your night vision by as much as 50 percent. This could make a walk or drive home at night very dangerous. So wear your sunglasses in the winter, and be careful.

R-E-S-C-U-E

Hundreds of fishermen, hunters, hikers and campers will get lost this year. Most of them will make it back okay, but an unlucky few will wander aimlessly until they freeze or starve to death. This does not have to happen.

If you get lost, think of the word RESCUE and what the letters stand for:

R Relax, never panic. Running around only uses up energy and further disorients you.

E Elevation. Find a hill or the nearest available higher ground. Don't aim for some promontory miles away. Just get higher so that you have a better view of the situation.

S Sit down. Use your map, compass, the sun's position and anything else at your disposal to get oriented. Most of the time, you will be able to right yourself. If you still can't figure out a way back, give thought to backtracking.

C Choose a course of action with a cool head. If you haven't reoriented yourself, build a simple shelter and make camp. If you think you know which way is out, be sure you can make it before sunset. If not, spend the night and rest.

U Utilize everything at your disposal. If you camp, collect enough firewood for the night. Make a comfortable bed and signal with your knife or mirror as long as the sun allows.

E Economize on your supplies. Drink water and eat, but save enough for a possible long stay. If your county has a search-and-rescue team, a smoky fire usually spells rescue in the morning. Save food for breakfast—you may have a long hike out.

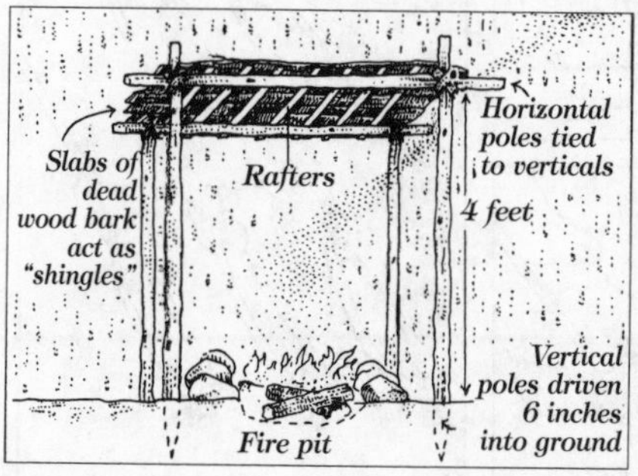

Foul-Weather Fires

Keeping a fire going in the rain is a challenge, unless you erect a shelter over the fire. Drive four 4 ½-foot-long poles into the ground about four to six inches beyond the fire. Connect them at the tops with horizontal crosspieces attached to the verticals with string or twine. Lay one-inch-wide rafters over the crosspieces to form a roof. With an ax, strip slabs of dry bark from fallen pine trees. Lay these shingles over the rafter poles to form a protective cover. It will shield your fire from most rains.

Foot Care

You should adjust the laces of your boots depending on the terrain, especially on a hill or mountain. At the start of a climb, loosen the toe and tighten the heel, which tends to shift and chafe in uphill travel. On the downslope, loosen the heel and tighten the toe. Experience will teach you when it is necessary.

This kind of adjustment can be done easily on boots that lace around hooks, but is possible on boots where the laces feed through D-rings or eyelets. The trick is to run the laces around each hook in a crossover loop. Boot laces set up this way won't slip; they stay tight where you want them tight and loose where you want to ease the pressure on your foot.

Know Your Half Hitches

A half hitch ***(Fig.1)*** is the simplest, and may hold against a steady pull, but it's better to go to two half hitches ***(Fig. 2)*** and finish the tie properly by pushing the hitches together and then snugging them against the object they're tied to by pulling on the standing part of the rope. (The standing part of the rope is the longer end—the end away from the knot.)

The timber hitch ***(Fig. 3)*** is useful for dragging a log or other heavy object. A simple hitch near the front end facilitates guidance ***(Fig. 4).***

The taut-line hitch ***(Fig. 5)*** is great for guy ropes

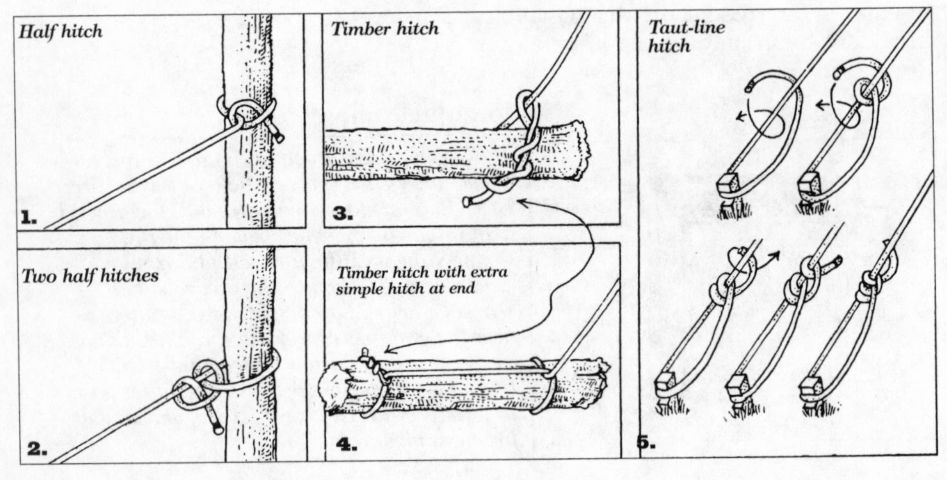

on tents. It can be tied on a taut rope. The important feature of this knot is that once it is tied, you can adjust the tension on the rope by pushing the hitch up or down. To tie the taut-line hitch, pass the rope around the tent peg and bring the end under and over the standing part and twice through the loop that's formed. Again bring the rope end up and under, over, and through the loop formed, and tighten.

With these few simple hitches, you will be able to quickly and securely attach a rope to almost any object for nearly any purpose.

Make Yourself Chiggerproof

Tough outdoorsmen are no match for chiggers (also known as red bugs, they're actually mites). Their four-stage cycle is egg, larva, nymph and adult. Eggs hatch into larvae and remain in this form until attaching to a warm-blooded host. Only after feeding on blood will they fall off.

Chiggers may sometimes carry disease, but most of the danger results from secondary infection caused by scratching.

To chiggerproof yourself before a field trip, treat your clothes with sulfur powder or insect repellent. You can also take sulfur cream-of-tartar tablets a few days before the outing.

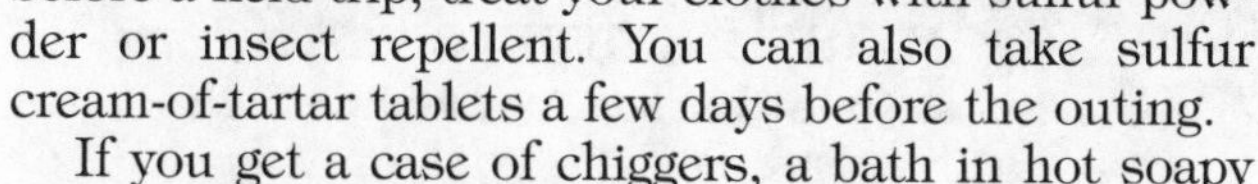

If you get a case of chiggers, a bath in hot soapy water with lots of scrubbing may wash them away.

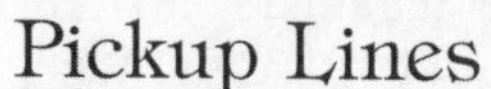

Pickup Lines

Pickup trucks are ideally constructed to be modified for hunting. Here are some things you might want to try:

* In the cargo box, install a pair of rear-facing seats, or a long bench seat or back-to-back seats. Hunting from a vehicle is generally illegal, but such seats are extremely useful when searching for lost hunters, transporting hunters to shooting areas, and scouting hunt country.

✳ Adding a roll bar provides a safety factor in the event of rollover, and the bar serves as a convenient body support for riders in the back. A few hunters attach a seat to the top of a roll bar for high-vantage scouting of antelope and other game in open country.

✳ A snowmobile ramp carried in the pickup box allows easy big-game loading. One man alone can load a huge animal if he has the advantage of a ramp.

Reading the Clouds

Before radio and television prognosticators came along to tell rural people how the weather was going to be, they had to figure it out for themselves.

Here are a few old ways of reading clouds:

✳ ***Delicate, soft-looking clouds foretell fine weather with moderate to high breezes.***

✳ A gloomy, dark, but very blue sky says windy but fair.

✳ ***Bright yellow clouds at sunset foretell wind; pale yellow means rain.***

✳ The softer the look of the clouds, the less wind there will be. Hard-looking clouds with rolled and ragged appearance tell of coming strong wind.

✳ ***Small ink-colored clouds mean rain.***

✳ Light clouds skimming across and beneath heavy masses will show wind and rain, or sometimes just wind, depending on the type of cloud cover that's above them.

✳ ***High clouds crossing in a direction different from the lower clouds foretell a change in wind direction.***

Map Practice

Whether you're a hunter, fisherman, hiker or wildlife watcher, it always pays to know where you're going. Here's a new way to let the information sink in. In addition to purposeful study with compass and pencil, tack one or more maps to the wall of your office or study. When you get a chance, take a few minutes to look over the map, and think of yourself taking various trails and routes. Do this several times per day, and by the end of the week, you've absorbed a lot of useful information.

The Body as Barometer

The next time you feel irritated enough to ram your fist through the wall of your tent or camper, take a look at the sky. Chances are bad weather is on the way, if it hasn't already arrived.

Dr. Michael DeSanctis of Minnesota's Buffalo Mental Health Center believes as much as one-third of the population is abnormally sensitive to weather changes. When it turns from fair to foul they can suffer lethargy, dizziness, headaches and depression.

Weather-sensitive individuals tend to have short-lived moods, low energy levels and low pain tolerance. You can put the blame on ions. Air ions affect the production of serotonin—a hormone that affects sleep cycles, sexual arousal and emotions.

DeSanctis claims negative ions, which are associated with pollution-free air, high altitudes and improving weather phases, can reduce anxiety, promote faster reaction times, and increase one's vigor.

Latitude Adjustment

If your outdoor trip will take you into the mountains and up a few thousand feet, consider the following when selecting clothing: It has been estimated that each 250 feet you ascend is equivalent to traveling northward a distance of almost 70 miles. So, if you live at an elevation of 1200 feet and plan to hike somewhere in the Rockies above 12,000 feet, you'll very likely experience a radical climate change. The equivalent would be moving from the warmth of Miami to the cold shores of James Bay in Canada.

Put strictly in terms of temperature, when the base of a 12,000-foot summit has a temperature of 85°F, the peak will likely be in the vicinity of 40°F.

Why You Should Watch Windchill

Outdoorsmen should keep in mind that crisp 30°F and 40°F days can change abruptly in late autumn, winter and early spring. According to the National Weather Service, a high wind can turn a 30°F day into a bone-chilling minus 5. This could kill if you are not prepared. The colder the actual temperature, the more severe the effect of wind. Before heading outside for a long time, consult your local weather channel and find out how cold it really is.

Watching Windchill

Wind Speed		**Temperature in °F**							
Calm		**30**	**25**	**20**	**15**	**10**	**5**	**0**	**-5**
3-6 knots	*5 mph*	*25*	*20*	*15*	*10*	*5*	*0*	*-5*	*-10*
4-10 knots	*10 mph*	*15*	*10*	*5*	*0*	*-10*	*-15*	*-20*	*-25*
11-15 knots	*15 mph*	*10*	*0*	*-5*	*-10*	*-20*	*-25*	*-30*	*-40*
16-19 knots	*20 mph*	*5*	*0*	*-10*	*-15*	*-25*	*-30*	*-35*	*-45*
20-23 knots	*25 mph*	*0*	*-5*	*-15*	*-20*	*-30*	*-35*	*-45*	*-50*
24-28 knots	*30 mph*	*0*	*-10*	*-20*	*-25*	*-30*	*-40*	*-50*	*-55*
29-32 knots	*35 mph*	*-5*	*-10*	*-20*	*-30*	*-35*	*-40*	*-50*	*-60*
33-36 knots	*40 mph*	*-5*	*-10*	*-20*	*-30*	*-35*	*-45*	*-55*	*-60*

The Single Knife

Hikers and hunters are economizing on the bulk and weight they carry. One of the items that should never be left behind (but one that often is) is a signal mirror.

If your knife blade is finely polished and wide enough, it can also be used as an emergency signal mirror. To aim it, stand and face the target you want to signal. Rotate the knife until the sun's reflection is on your feet. Slowly turn the knife, so the reflection moves from your feet and toward the target.

One of the drawbacks to using a mirror is that you can only signal targets between you and the sun. One way to get around this is to bounce the light from your knife onto another mirror or knife. This method allows you to signal in any direction, regardless of where the sun is positioned.

Don't Doze on Your Stand

According to numbers compiled by the tree-stand industry, stands accounted for 17 deaths, 30 permanently disabling injuries and an estimated 600 lesser but still serious injuries in 1992. The most common problem is falling while ascending or descending a tree. Dozing off and plummeting is another frequent cause, giving the expression "falling asleep" a new and unpleasant meaning. Cold, muscle stiffness, fatigue and mechanical failure of the stand all contribute to falls. The prevention: Inspect your equipment before entrusting your life to it; climb slowly and carefully; always wear a safety strap or harness when climbing to, sitting in, and leaving a stand.

Cool Hands

To roast smaller pieces of meat or fish safely, take along a few extra paper plates and impale them on roasting forks close to the handles to protect your hands from the heat. Remember, though: They are paper, so make sure they don't get too close to the flames.

Navigator Tips

Still-hunters tend to be among the most skilled of deer hunters and also among the most bold, venturing into places where others are too lazy or too afraid to go. It's probably also true that still-hunters get lost more than their stand-hunting compatriots. When you're moving along, especially in new country, concentrating intensely on finding deer, it's not difficult to suddenly look up and find yourself a bit turned around. Nothing serious, but a situation worthy of some concern. Usually a little calm reflection settles the matter.

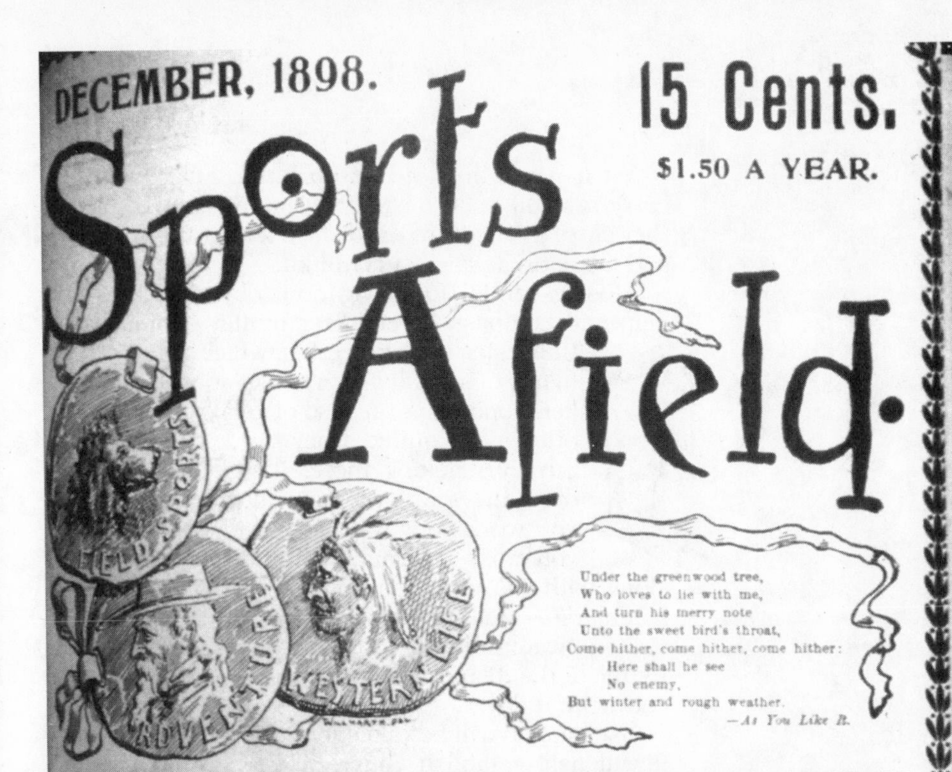

DECEMBER, 1898.
15 Cents.
$1.50 A YEAR.
Sports Afield.
FIELD SPORTS
ADVENTURE
WESTERN LIFE
Under the greenwood tree,
Who loves to lie with me,
And turn his merry note
Unto the sweet bird's throat,
Come hither, come hither, come hither:
Here shall he see
No enemy,
But winter and rough weather.
—As You Like It.

THE ELIXIR OF LIFE.—Our Camp in the Montana Highlands.

Yet it never hurts to be prepared. Smart hunters carry a small pack on their backs or hips in which they carry a rolled-up parka shell, a map and compass and a comprehensive survival kit.

Most essential for navigation, obviously, are the map and compass. Every deer hunter should learn the basics of orienteering, which are available in any of several dozen books now on the market. Topo maps are also of great value and can often be acquired from the U.S. Geological Survey. Lacking these, it's often helpful to draw up your own map, however crude, as you work your way through new country. If you cross a stream, mark it down on the map, along with the direction you took from it. Obvious landmarks—a clump of standing dead trees, a rocky cliff, a waterfall—should also be marked down. Estimate the distance between each recorded feature and write it down.

Such a map will be valuable if you become lost, and it will help establish your sense of the terrain and increase your prowess as a hunter. As a bonus, it will serve as a permanent record, should you want to return to a particular spot again.

Hot Water, Cold Feet

If you have camped at high altitudes, you have very likely slept in an icy sleeping bag. One way to stay warm is to include a flannelette-covered hot-water bottle among your gear. The bottle makes a fine heater in the bottom of your bag.

Why Life's an Itch

Hunters returning home from the field sometimes take with them a maddening, itchy rash, yet they can't remember rubbing up against any poison ivy, poison sumac or poison oak. They may not have, for North America is home to a wide variety of other plants that can cause contact dermatitis.

One of the worst is virgin's bower, a memeber of the buttercup family found in most of the eastern U.S. If the fresh sap comes into contact with your skin, expect a nasty rash with blisters and ulcerations. Creeping buttercup, a relative of virgin's bower that produces yellow flowers, contains similar toxins.

Most of North America is home to the stinging nettle, a plant with tiny, sharp hairs that have something akin to a bladder located at the base. The bladder contains a mixture of at least four toxins, including histamine and formic acid. When a person touches them, the nettle hairs pierce and release poison into the skin. The effects can last for more than an hour.

Washing with warm water and soap helps if done soon enough. Use calamine lotion when it's too late.

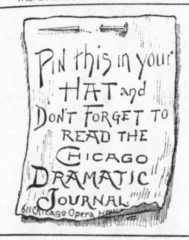

7
Essential Gear

The mental equipment which one totes in his upper story is really of greater importance than the material outfit which he drags on the toboggan.

—W. Dustin White, *Sports Afield*, February 1916.

A Still-Hunter's Checklist

Wearing the right clothing can improve anyone's chances of scoring while still-hunting for deer. Any crinkly, slick or hard-surfaced material is inappropriate. It simply makes too much noise when you move and is particularly loud when you rub against brush or a tree branch.

Instead, wear either cotton, wool or one of the new fleece-type synthetic clothing materials. If the law requires, or if other hunters are in the area, wear blaze orange in the form of a hat and vest, for safety's sake. If you're in an area by yourself where no other hunters are present and the law does not require orange, wear camouflage clothing.

Shoes should be either moccasins, tennis shoes or soft leather boots, preferably with neoprene or crepe soles, which tend to be quieter than lug soles.

Other items you'll need for still-hunting are a canteen or wineskin filled with water or juice, flashlight, knife, drag rope, compass, topo map, license and tags, lunch or snacks. Also bring waterproof matches in a waterproof container or a butane lighter and space blanket in case you get temporarily lost. Binoculars in the 6X to 8X range are perfect for still-hunting.

Short-range shooting is most common when still-hunting; thus calibers such as the 35 or 30-30 can be effective in the majority of situations. However, there's always the chance you'll spot a buck 225 yards across a field or on an opposite ridge in the

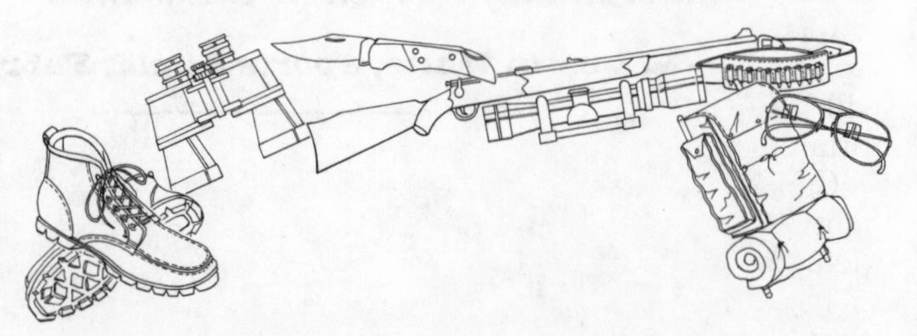

mountains. In those situations you'll wish you'd opted for a flatter-shooting caliber. Some of the best calibers for still-hunting include the 270, 280, 308, and 30-06. Bullet weights of 150 to 180 grains are ample for deer-size game with these rifles. Scopes should be low power such as 1.5X, 2X or 4X, since most shots will be fairly close.

Bungees for Deer

Elastic and/or rubber tie-downs (bungee cords) are ubiquitous today, and every hardware store displays them in assorted lengths and colors. Sportsmen have always found a myriad of uses for them, so it is no surprise that the tie-down is extremely helpful with the ladder stand. Most of these stands are secured to a tree with either a nylon strap or a chain, and additional tethering with a couple of long, stout tie-downs makes the ladder a much more secure perch.

A tie-down around a tree can anchor hooks that will hold binoculars, daypack, bow, etc. Also, a couple of tie-downs can securely fasten the folded ladder for transporting into and out of the woods.

Blood Damages Bluing

When blood comes into contact with gun bluing, it can cause significant damage. The acids in blood can remove the coating, exposing the metal surface and causing rust spots and surface pitting. Since dried blood on a blued surface is sometimes difficult to see, preventive measures are necessary. Keep your gun away from the gutting and dressing operation. Upon returning to camp, carefully inspect your guns, wipe them clean and dry, then oil them.

Gear for Hollows Hunting

Hollows hunters have to be patient. Deer will travel hollows all day long, so you have to be prepared to watch and wait the whole time you're afield.

Wear the least amount of clothing you can get away with on the trek to your stand. When you get to your stand, unload the rest of your clothing from a daypack, and as you cool off, gradually put on additional clothes. Though it is a bit of a nuisance, you can stay much warmer this way.

When you first reach your stand, it is a good idea to change your socks. Put the wet socks in a plastic bag and stash them in your daypack. Two pairs of socks are a must for cold weather.

Carry a Thermos of hot coffee, soup or cocoa, and sip it slowly at intervals throughout the morning. It will raise your core temperature and keep you comfortable longer.

Carry good raingear. There are several excellent lightweight varieties on the market today. Be sure to include a waterproof case for your gun. Also, be sure to include binoculars in your pack. The minis are great. While you're sitting on your stand, take the time to carefully scan the area for deer and other wildlife. You'll be surprised at how much faster the time will go.

Hollows stay darker later in the morning and get darker earlier in the evening, so a good light-gathering scope is a necessity. A 2X to 7X variable is a good choice, especially if your hollow has some distance across it. The lower power comes in handy for sorting out holes in brush and also for the close shots that occur frequently.

Be sure your rifle has a sling; using one greatly steadies your shots.

Remember to stay out there all day. It takes only a minute for the buck of a lifetime to cross from the next ridgev . . . and trot right into your sights!

A Tool From the Past

Splitting the pelvic bone of a large game animal and quartering the carcass can be difficult with just a knife, so many outdoorsmen carry a relatively heavy hatchet or collapsible saw for this task. Another option is the tomahawk, a light and efficient tool. Its small head renders it less-than-ideal for splitting firewood and similar tasks in camp, yet it is a perfect choice for quartering big game.

Modern reproductions of the colonial ax, weighing less than a pound, are available for around $20. Their long handles give them the leverage necessary to sever bones and to cut the poles used to spread a rib cage for cooling. Because tomahawks are made from mild steel, they are easy to sharpen with a file and tend to resist chipping under contact with bone.

To carry a tomahawk safely and comfortably, cover the edge with duct tape and slip it under the cartridge belt worn outside your hunting coat. The bottom curve of the head rests perfectly over the belt and keeps the covered edge away from the body.

Tomahawks are available in most catalogs that cater to blackpowder shooters and rendezvous participants.

Easy Tree Stands

If you intend to use a tree stand on public land or on private land you don't own yourself, it is probably illegal to drive nails into trees to build a stand. But don't despair. Dozens of manufacturers sell excellent portable tree stands that are safe and easy to install.

Non-Commercial Scents That Work

Though there are many excellent commercial scents on the market, some hunters prefer to use natural scents that they find in the field or at home. Here are some popular scents you won't find in sporting-goods stores.

✻ ***Pine-Scented Disinfectant:*** One of the most successful bowhunters in the country is Ben Rodgers Lee of wild-turkey hunting fame. When hunting deer in pine woodlands, Rodgers washes his hunting clothing in pine-scented disinfectants. These products are usually used around the home for cleaning cabinets and washing garbage pails and such, but they can also be used for laundry. Their high pine-oil content leaves clothing smelling like the woods. Rodgers takes a high number of whitetails with his bow each year, so it must work.

✳ ***Pine and Cedar Needles:*** You can acquire the scent of pine or cedar woodlands simply by pulling a handful of fresh needles, breaking and bruising them and wiping the juice on your hunting clothes. This is perhaps the most convenient scent to use when you've forgotten a commercial scent.

✳ ***Rabbit Tobacco Weed:*** The weed commonly called "rabbit tobacco" (*Gnaphalium obtusifolium*) is found throughout most of the white-tailed deer range, and pioneer hunters discovered that it made a good masking scent. They would pull a handful of the strong-smelling silver-gray leaves in the fall, bruise and wad them into a tight ball and place them in a coat pocket or hat brim.

You can find rabbit tobacco along dry, open, weedy areas such as old fields, logging roads and abandoned homesites. The plant blossoms from August to October with a white-to-cream-colored flower. Each stem rising from one to three feet above the ground will have many flowers on the end. The leaves are narrow and usually dark green on the top and silver-white underneath.

✳ ***Apples:*** In apple country you can cut apples in half and rub the juicy side on your pant legs. A better method is to find some pure cider, carry it in a small plastic squirt bottle and use it on pant legs and trees near your stand.

✳ ***Turpentine:*** This is another good masking scent to use when hunting in piney woods. It's a pure product of the pine tree, and a few drops on the boots, pant legs or on trees around your stand will do the job. Don't use too much or you'll make the deer extremely cautious.

✳ ***Cattle Scent:*** Deer that range around dairy or beef cattle and share feed and salt with them are accustomed to the smell of the cattle. There are two ways to get this scent on your hunting clothes. One is

Scope Adjustments

Your scope needs adjusting and you haven't got a screwdriver? A coin, either a penny, dime or nickel, works better than most screwdrivers. Screwdriver blades are generally much too narrow and can cause unnecessary wear. Coins, on the other hand, provide you with a much better fit, and best of all, you are seldom without one.

to hang your clothes in the barn overnight. The other is either to walk through fresh cow manure and get it worked well into the crevices of the boots or to smear fresh manure on trouser legs.

✳ ***Deer Urine:*** For obvious reasons, this is one of the best masking scents. It is also good lure to use during the rut, especially when hunting scrapes.

You can supply yourself with deer urine simply by removing a deer's bladder while field-dressing the animal—being careful not to spill the contents on the meat—and emptying the urine into a plastic jar or bag. Store it in your freezer until your next hunting trip. Be sure to mark the bag or jar well. If your wife won't let you store it in the freezer, you can preserve it by mixing one ounce of benzyl benzoate to each pint of deer urine. This mixture will keep without refrigeration.

If you are a stalker or stand-hunter, you will probably want to cover your scent when walking to your stand or stalking. Deer urine is the most effective means of doing this. Place a few drops around where your boots' uppers meet the sole. Also put a few drops around the bottom of your trousers.

It can be easily carried in a small medicine bottle with a squeeze-dropper top.

Smell Like a Cat

Bobcat urine, which can be purchased from most trapping supply houses, is a scent deer smell all the time. Since this is a strong scent, use only a few drops at a time.

Sled Your Game

A simple plastic sled can turn the chore of moving big game into child's play. Two types of children's sleds are suitable for towing game, and both are inexpensive. A molded-plastic sled has hold-on ropes in the right places for tying down a heavy load and sides high enough to keep the game from being dumped out on bumpy ground.

Also suitable is a flat sheet of plastic that rolls up (much like an Ensolite pad, only thinner). These super-lightweight sleds are particularly suited for packing trips.

However, tying down a load on the flat plastic is difficult, and ropes usually end up being dragged underneath the surface, causing friction. Try cutting holes in the sled's sides and curling them together around the load; then attach a lead rope for easier towing.

Both types of sleds tend to disintegrate quickly with hard use, but will generally last for a hunting season or two. And it's much easier than packing it all on your back in five or six trips.

Strop for a Razor Edge

A good sharpening stone will put a fine enough edge on a knife used for meat-cutting purposes. A better edge for skinning game, however, is a razor edge, and it can be obtained by stropping the blade, barber style.

You'll need a ¾-inch-thick block of wood; a can of Clover valve-grinding compound, which can be purchased at most automotive supply stores; and a piece of heavy leather such as that used in saddles or belts.

Cut the wood and leather to the same size—2 ½ x 8 inches is handy and efficient. Then glue the leather, rough or back side up, to the wood with epoxy glue. For security, use six small screws, countersunk into the leather.

Use a putty knife to rub a generous amount of the valve-grinding compound down into the leather; let dry overnight.

Now hone your knife as usual, then strop for a razor edge.

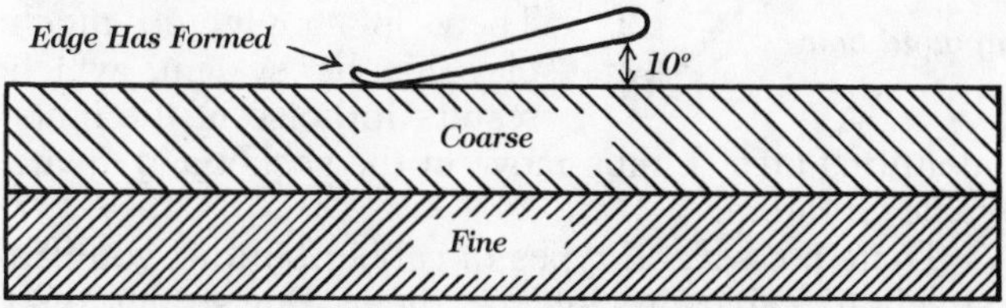

For razor-sharpness, stroke the blade on a stone—moving from coarse grain to fine—then strop on heavy leather.

Rust-Free Knives

Never store a quality knife in its scabbard. Leather cured with tannic acid will, over time, draw moisture and stain even the finest steel, no matter how well oiled. At the end of the season, clean your knife thoroughly, removing all the dirt and residue. Then rub on a light coat of Vaseline. Wrap the knife in waxed paper and store in a dry place. The scabbard can be cleaned and rubbed well with neat's foot oil and stored separately.

The Quiet Bowhunter

In the early days of compound bows, the limbs—seated in the mounting in the handle riser—were notorious for making a creaking noise. Squeaking wheels have also been a problem. The solution for the former was to draw it once when you got on stand, just as a check. A can of baby powder usually solved the latter. There's a lubricant available today with microscopic beads of Teflon in it; it works well.

Other things to check: cables slapping on yokes, on limbs and on each other; squeaking in the cable guard; and cables rubbing in the eccentrics or cams because they're not properly aligned.

There's also a "high-performance slap," most often found in the cam bows. So much energy goes into a shot with such a bow that the bow sort of

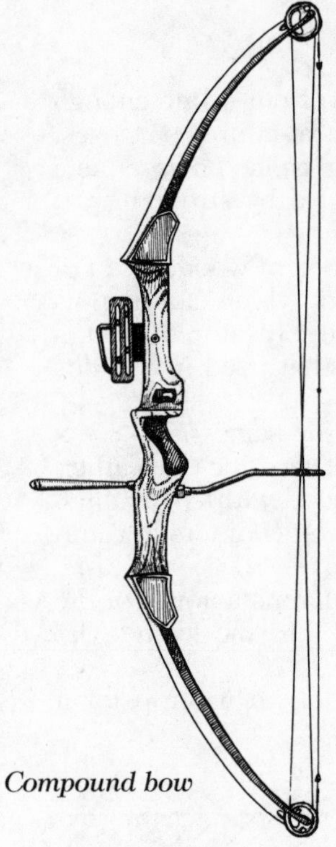
Compound bow

shakes the way a dog shakes a squirrel. It more or less comes with the territory, but some manufacturers are backing off the performance in order to solve the problem. The bow will give up 10 to 12 feet per second in arrow speed, but it will be much smoother and quieter.

High-performance cam bows have created a noise problem also with arrows. The standard arrow matched to the peak draw-weight is too light; a heavier arrow is needed to absorb the energy and silence the bow.

Sometimes a bow will be noisy because the limbs are not under enough tension. There is simply too much play in the system, which results in noise in the limb-mounting area. Limbs move in the socket and create cable vibration.

There are several types of rubber buttons you can attach to cables to silence them. Bowstrings also need silencers of some sort. The spidery rubber ones are good because they don't absorb moisture and they last a decent amount of time.

Arrows can rattle in a quiver when the rubber clips are too loose; the arrows slide down a bit and become free riding in the quiver hood. Clip-and-arrow fit can work the other way, too.

Fletching can rattle or scrape against other fletching. This can be solved by carefully placing the arrows in the quiver so fletching won't strike other fletching, i.e., to spread those arrows enough to keep them from rattling. On many quivers, the hood and clip are too close together, allowing play in the nock end of the arrow. A longer distance between hood and clip would hold them more rigidly.

A good quiver hood will have enough room to allow proper placement of a broadhead. Some of the newer broadheads are wider than old styles, and can jam together in a smallish quiver hood. This leads to scraping and rattling against other heads and against the hood itself. It also leads to dull blades.

A foam panel in the hood will often hold broadheads best when the quiver is new, but that foam can be cut to pieces over time and create rattling room. For this reason, quivers with individual hard rubber cups for broadhead tips are probably the best way to go.

A rubber gasket–type piece placed between the bow and the quiver, or the bowsight mount and the quiver, often helps silence the entire unit. More than one quiver—otherwise carefully silenced—has spooked a deer because the hunter forgot to check the attachment and screws.

Bowsight pins have a way of working loose, especially during an auto trip. Check them before you head into the woods.

Don't let a big whitetail get away because of a sudden rattle in your bow.

How to Recycle Broken Arrows

Every bowhunter who uses aluminum arrows occasionally finds himself with bent or broken shafts. Here's a way to recycle them: Cut the damaged shaft into three-inch lengths with a hacksaw. Attach two pieces to the front of a bow with tape and several more to the framework of your portable tree stand.

When afield, gather leafy branches—12-inch lengths of bushy pine boughs are best—and insert them into the open ends of the old arrow shafts. Presto: frontal camouflage.

Gear Smarts

Pick your hunting kit with care. Let it evolve. Don't be lured in by every claim of miracles. Conversely, don't shun new developments just because they are new. Spend what you have to. In two weeks you will forget the price; in two years you will remember the quality. More goods are overpriced at $19.95 than at $79.95. And remember that if it were some widget that made a good hunter superior to the rest of us, being a good hunter wouldn't mean much.

Keep Your Stand Private

Deer hunters who use permanent or portable tree stands are often concerned about theft of their equipment or finding their permanent stand occupied by someone else. Removable screw-in tree steps are available, but here's an inexpensive method to convert your existing steps to removable ones. Simply grind off or file off the pin holding the step to the worm screw and replace it with a hardened hinge pin, which is available at any hardware store. Do this with the four bottom steps on the tree, both for safety reasons and so that any would-be thief has to hug the tree to get to your stand. When leaving the stand, remove the pin and step, leaving only the worm screw in the tree. As always, make sure screw-in steps are legal in your area and periodically inspect the steps and stand for any deterioration.

Components of a Survival Kit

Almost all seasoned backcountry travelers carry a small survival kit with them at all times. In it should be the items necessary for signaling, shelter, fire-making, navigation and first aid. It packs in an army surplus individual first-aid pouch, weighs two pounds and fits easily on your belt or in your daypack, backpack, vehicle glove box or tacklebox. The kit is composed of the following items:

✻ Blaze-orange smoke signal for daytime signaling in conjunction with ground-to-air signals.

✻ Change for a telephone call. Many stories are told of hunters who made it to a pay phone and had no change for a call.

✻ A coil of 20-pound-test fishing line to be used

for making a shelter, mending clothes, and making simple snares.

✳ Space blanket that can be made into a lean-to, used as a blanket (it is excellent at reflecting body heat) or used for signaling.

✳ Aspirin.

✳ Band-Aids.

✳ Police whistle for signaling.

✳ A candle stub to use as a fire starter.

✳ A Boy Scout-type pocketknife.

✳ Antiseptic for scratches and wounds.

✳ A signal mirror.

✳ Powdered beef broth or individually wrapped bouillon cubes. They have some nutritional value and make wild food dishes taste much better.

✳ Wallet-size survival guide.

✳ Waterproof matches.

✳ Water-purification tablets.

✳ A small tacklebox made by winding several feet of 6-pound-test fishing line around a plastic pill bottle and holding it in place with tape. In the bottle are several small hooks, split-shot, a small panfish popping bug and a small dry fly.

✳ Tweezers for removing splinters, etc.

✳ Lip protection from wind and sun.

✳ Metal match to be used as a back-up fire starter.

✳ Size 000 steel wool to use with the metal match to get tinder started. One spark in loose steel wool, even when wet, produces a hot glow that will start dry tinder.

✳ A back-up compass.

Take Care of Your Tent

If you are a deer hunter who spends a lot of time camping out, remember that a tent will only serve you well if you hold up your end of the bargain, and treat it with a great deal of care. If the tent is well-made, it should last you many seasons.

✳ ***Muddy Terrain:*** When camping in wet weather, put a plastic bag over your boots before you bring them in. It will protect the floor of your tent and help you keep your other gear clean. Or, if you haven't yet bought a tent, look for one with a fly that extends past the doorway, so you can keep your boots dry, but outside.

✳ ***Packing Poles:*** Carry your tent poles and ground stakes in a separate stuffsack to prevent them from poking holes in your tent.

Mask Your Scent

Bowhunters were quick to realize the value of products used to mask human scent and have come up with dozens of ways to apply them. One of the most effective methods is to place all the clothes you will be wearing—from socks and underwear to camouflage gear—in a large plastic trash bag. Then simply dribble some masking scent over them and seal the bag with a wire twist-on. Let the clothes sit overnight, and the scent will permeate all, greatly reducing a hunter's chances of getting winded by game.

✳ ***Tent Storage:*** No matter how tired you are after a trip, never store your tent right away. Moisture from rain or condensation can rot the fabric. Set the tent up in a dry place after cleaning it with a damp cloth and let it dry completely.

✳ ***Sun Exposure:*** Avoid setting up your tent in an exposed area for too long. Ultraviolet rays have been proven to damage a tent fly.

No-Slip Soles

When you finally bring out last year's hunting gear, you may find that the felt liners or innersoles from your winter boots are worn and need replacement. Don't throw them out.

Cut the liner to match the sole of an old sneaker. Fasten the felt to the sneaker sole with a nonsoluble glue. Place a heavy object (a brick works well) on top of the sneaker overnight, and by morning you have an all-purpose non-slip shoe.

Wear with stocking waders for stream-fishing or alone for safe walking on rock jetties. High-tops are best for deer hunters, as they provide extra protection against ankle scrapes from sharp rocks. Another advantage of canvas sneakers is that they dry quickly.

If you glue felt bottoms to the soles of an old pair of rubber boots, reinforce them by wrapping duct tape around the sole and over the top of the foot, as the grooves on the boot bottoms make it difficult for the glue to hold.

Pliable Parachute Cord

Never leave home or camp without a hefty coil—maybe 50 feet—of parachute suspension line. This is strong, versatile stuff for lashing and pulling and hoisting. A couple of strands will do as much as any rope. The inner core can be pulled apart, yielding seven thin filaments suitable for the most delicate work. Combined with four feet of wide nylon webbing to use as a sling over the shoulder, suspension line will serve well as a drag. Remember to melt the ends with a lighter to prevent fraying.

A Bargain Meat Saw

For deer hunters on a tight budget, a bargain-priced meat saw can be assembled by using a light, sturdy hacksaw (purchased at a garage sale if you don't have one) and adding a small meat-saw blade (these are widely available at hardware stores in various lengths). The blade can be easily tailored to the hacksaw's size by cutting and drilling.

When you're traveling in the woods, the saw should be assembled with the teeth facing inward and covered with masking tape to prevent abrasion against other items in your day-pack. Wrap the saw with a 10-yard length of nylon cord to keep it from making noise. The nylon cord also comes in very handy when you're dealing with a large mule deer: Use it to hold the animal in place during the initial gutting and skinning operations.

Free Scent Cover

While there are many formulas on the market that will allow bowhunters to smell like the woods, all you need is a plastic garbage bag and a shovel for digging up organic material from around your stand. It makes the perfect masking scent.

Begin by washing your hunting clothes as you normally would, substituting borax or baking soda for detergent. Hang the clothes out to dry, or stick them in the dryer—without any scented softener.

Then get out the plastic bag containing your new deodorant. The fresh topsoil under the leaf litter makes the best de-scenter, as do the leaves, twigs and grasses from around the site. Put the clothes, including your boots if possible, in the bag on top of all the collected material (a few inches in the bottom of the bag is enough). Tie off the bag and set it aside for one or two days before your next hunt.

Lighting Your Way in the Woods

Most deer hunters make their way to their stands in the early morning dark, so it's a good idea to carry along a penlight. It will provide enough light to let you safely step over logs, stumps and rocks and avoid brush. And because it is so dim, the light won't alarm deer. For safety reasons, a penlight should be standard equipment for anyone hunting in the woods at dawn or dusk.

Mule Deer Optics

Good optics are among the most important items you'll need for late-season mule deer hunting. Using open sights on a rifle is simply out of the question in this vast, sprawling country. You'll need a minimum 4X fixed scope or, better yet, a variable in the 1.5X to 6X, 2X to 7X or 3X to 9X range.

Besides a quality scope, binoculars are a must. Choose 9X or 10X for the best magnification. If you already own a pair of 8X binoculars, however, they'll perform adequately.

A final piece of glass you should have is a spotting scope in 30X or 40X, or a variable covering this range. These will save long, needless stalks by allowing you to thoroughly check out a buck's headgear from great distances. They can also enable you to spot faraway game that binoculars or the naked eye could never pick out.

Where to Find Good Wool

Quality woolen clothing suitable for deer hunting is relatively scarce in many towns, compared to synthetic garments. The availability problem is one reason more hunters do not use wool. Try outdoor specialty shops that cater to backpackers, skiers and climbers. Generally, the best hunting and fishing supply outlets in the northern and western states and in Canada carry more of a selection of wool garments suitable for hunting. If that does not work, shop by mail. Outdoors catalogs that offer good selections of wool garments generally back up their products with exchange or money-back guarantees. Here are some things to look for:

✱ Military-surplus wool pants and shirts are a good buy for the money. The style may be outdated, but the wool works fine.

✱ Try not to be shocked by the price of some of the fine, tightly woven wool shirts, jackets, pants and caps. In most cases, the investment will be a long-term one.

✱ Most fine woolens can be hand-washed in Woolite or dry-cleaned. The going rate for dry-cleaning a wool hunting shirt is about $6.

✱ Avoid buying wool garments lined with materials that negate the breathability and comfort of wool.

✱ Look for the "100% Virgin Wool" label or other such "pure wool" logo.

Wet-Weather Clothing

Wearing the proper clothing for bad-weather deer hunting is vital. You can't hunt effectively if you're wet and cold, and the danger of hypothermia is particularly high during inclement conditions.

Boot pacs are best in extreme weather. In milder conditions, totally waterproof leather or rubber boots will suffice. Wear at least two layers of socks—first a pair of Thermax or polypropylene, then a pair of wool.

Your long underwear should also be made of either Thermax or polypropylene, in order to wick moisture away from your body. Then put on one layer or more of cotton, wool or synthetic shirts and pants, topped off in bitter weather by a down- or synthetic-insulated coat or vest. If it's wet out, add Gore-Tex or a similar type of raingear as your final outer layer, plus a warm waterproof hat and gloves.

If you plan to still-hunt in wet weather, consider one of the Gore-Tex coats with wool outer layers for added quietness, such as those sold by Columbia or Woolrich. You can also make your current Gore-Tex outerwear quiet by donning loose-fitting cotton or wool clothing on top of the Gore-Tex.

Monocular From Broken Binoculars

A very serviceable monocular can be salvaged from broken binoculars. The compact monocular can be carried with a neck strap or in one of your pockets. Use a fine-tooth hacksaw to separate the halves of broken or no-longer-useful binoculars. Buff the edges, and it's ready to go. Binoculars with eyepieces that can be adjusted individually provide two monoculars. Also note that optical shops often have damaged binoculars at a near-giveaway price.

Camo Cover for Your Gun

Some hunters paint their guns and other use camo tape. If the thought of subjecting a prized weapon to such treatment makes you shudder, then try this alternative camouflaging method.

Make a camo cover for your gun. Simply lay the gun on a piece of camouflage material (which can be purchased at most surplus stores), and trace its outline with tailor's chalk or a felt-tipped pen. Flip the gun over and trace once more. Cut out this tracing, leaving a one-half-inch margin outside of the chalk line. Now, with the cloth turned inside-out, stitch the edges together; leave the ends open for the muzzle and the butt plate. Make cuts to accommodate the trigger guard and action. Then all you have to do is turn the material right-side-out and slip the case over the gun.

If you hunt in areas of frequent snow cover, make a slip-cover for your gun from an old sheet.

A Backcountry Hunter's Checklist

Comfort Items:

- shelter (lean-to or dome tent)
- one-eighth-inch-diameter nylon rope
- sleeping bag
- mattress
- ground cloth
- towel, washcloth and soap
- toothbrush and toothpaste
- toilet paper
- coat, vest and parka
- cap
- down booties
- underwear, socks and handkerchiefs
- gloves or mittens

Eating and Cooking:

- food

* stove (optional)
* fuel (optional)
* cup
* spoon
* 1 ½-quart pot
* dishtowel, dishcloth and detergent

Hunting Gear:

* hunting licenses and tags
* rifle
* ammo (20 rounds)
* knife
* knife sharpener
* game saw
* ax (optional)
* game bags
* binoculars
* spotting scope (optional)
* sunglasses
* maps

Safety Items:

* matches and cigarette lighter
* fire starter (Sterno)
* flares and signal mirror
* compass
* small first-aid kit
* flashlight
* extra bulb and batteries

Hey, Turkey!

No woodsman should ever be without a diaphragm turkey call. With it I can squeak at coyotes, bugle at elk, cluck with turkeys, honk at geese, and mimic any of those sounds to get the attention of hunting partners without yelling "Hey, Louie!"

How to Buy Binocs

When choosing binoculars for field use, don't buy more power than you need. The higher the magnification, the more difficult it becomes to keep the image steady without using a rigid support. High-power binocs tire eyes quickly. Eight power is the upper limit for most hunters, and is usually more than enough. Zoom models let you pick the power you actually need. The trade-off for that flexibility is that zoom glasses don't transmit as much light as fixed-power binocs.

Traditional porro-prism glasses have wide fields of view and excellent light-gathering ability. However, they are also bulky. If a wide view is important, or if the binocs will be used extensively at night, porro-prisms make sense.

Like zoom glasses, armor-clad and waterproof

binocs have their niches. The trade-off here is added bulk: Do you really need these special features? Spend five minutes actually looking through an intended purchase. Some less-expensive binocs are much better than others. Look first to see if what you focus on is sharp and bright. If so, keep looking. If the lenses are misaligned, they may cause you to develop a headache.

Eyeglass wearers should first try new binocs without them. The focus adjustment is often sufficient to correct vision.

The Two-Knife Hunter

Hunters need two knives for a cleaner, faster job of field-dressing, with minimal chance of punctured entrails, tainted meat and nicked hides.

Handiest is a small dropped-point or clipped-point knife to make the initial ripping cuts and, later, to open the stomach cavity precisely and loosen the lower bowel around the vent. A blade length of 2 ¾ to 3 ½ inches is quite adequate.

For skinning and for heavier cutting chores, switch to a larger blade. It can be anywhere from four to six inches, depending on your tastes and the size of the animal. Blade contour should be deep bellied with lots of curved edge. Because it is part of a two-knife set, you can choose a profile that is upswept at the point or, if you prefer, straight-backed.

Get Ready for Deer Season Now

Summer may seem an odd time to think about deer hunting, but it's not. Put off your preparations, and things will be in shambles by opening day.

Guns and bows will lie gathering dust. Muscles will be out of shape. Nonresident and controlled-hunt license deadlines will have passed. Gear will need repair. The cabin roof will still leak. And if you're really lucky, the land you've always counted on for a productive opening day will have been sold and posted.

Granted, this is a worst-case scenario. But few of us can wake up on the opener after a long off-season and expect to find everything primed and ready for another fall. Here are some of the most important preparations you should make during the summer months.

*Confirm that the property you hunt is still held by

the same landowner and that he will still let you hunt.

✱Apply for any licenses, controlled-hunt area permits or other paperwork that must be taken care of before certain cutoff dates.

✱Start an exercise program. Gentle or severe, the choice is yours, but nearly everyone needs to get in better shape before going deer hunting.

✱Book a guide if you plan to go on an outfitted hunt. The best guides fill early and may, in fact, already be booked on the days you want to hunt. If so, reserve a date for next year.

✱Clean your rifle, muzzleloader or shotgun; tighten scope mount, sight and quiver screws; tune your bow; check for string wear.

✱Check your clothing, tree stand, packs, rattling antlers, grunt tubes, scent supply, binoculars, knife, flashlight and any other gear you might need for repairs or replacement.

✱Check your ammunition, arrow and broadhead supply.

✱Request time off from work for the dates you wish to hunt.

✱Line up lodging, unless you plan to camp.

✱Scout your hunting area for paths and runways used by deer.

Scopes for Deep Timber

In deep timber, where you often come up on deer suddenly, and at fairly close ranges, 4X scopes are as powerful as you'll need. If you use a variable-power scope, turn it to the lowest setting before you start to hunt. You won't have time to change the setting when you see a deer; and if your scope is in the 9X position, all you'll see when you bring the rifle up will be an eyeball full of blurred timber.

Ordering Topo Maps

Among the most crucial items for a successful deer hunt are quality maps. Topographic maps show the contours of the land, natural funnel areas for deer, clearings, streams, steep escape areas, swamps, old homesites, roads, streams and other useful details about your hunting area. They enable you to mark down the locations of sign you find during scouting trips. And finally, topos allow you to discuss hunting strategy with friends.

It may take several weeks to receive maps by mail, so be sure to order them well in advance. The most detailed topographic maps available—and the ones of greatest use to the deer hunter—are the 7 ½-minute quadrangle maps that use a 1:24,000 scale: One inch on the map equals 24,000 inches on the ground, or about 2000 feet. To order specific maps for the area you want to hunt, contact the U.S. Geological Survey at 1-800/USA-MAPS (see address in Appendix, page 247), and ask for a free map-ordering brochure and state index. Then order the topos

H. J. UPTHEGROVE,

VALPARAISO, INDIANA.

Sportsmen's and Civil Engineers'

CLOTHING

Illustrated Catalogue with samples of qualities and shades of Corduroy, Mackintosh, Canvas, Flannel, etc., mailed free.

Agents: Simmons Hardware Company, St. Louis, Mo.

Chicago Agents: Von Lengerke & Antoine, 277 Wabash Avenue.

OF INTEREST TO SPORTSMEN

is the

Southern SR Railway

HUNTING - AND - FISHING - BOOK

Descriptive of the best localities in the South for various kinds of game and fish. It gives the Game Laws of all the Southern States. Send 2 cents in postage for copy of "HUNTING AND FISHING IN THE SOUTH" to

J.M. CULP, Traffic Manager, Washington, D. C.
WM. H. TAYLOE, A. G. P. A., Louisville, Ky.
WM. A. TURK, G. P. A., Washington, D. C.
C. A. BAIRD, T. P. A., Louisville, Ky.
J. C. BEAM, JR., 80 Adams Street, Chicago, Illinois.

you need. You can also ask for brochures that will help you learn basic map-reading skills.

Have Gun, Will Try to Travel

Getting there is half the fun, and twice the headache, especially if you're traveling with firearms. Here are some ways to make matters less painful and time-consuming.

✳***Before You Go:*** Work out details beforehand, rather than trying to untangle snarls when they arise. First, learn the laws governing the importation of firearms into the locale you're traveling to, and don't forget that some U.S. states and municipalities may have laws affecting what types of firearms you may possess there. Your outfitter, booking agent or travel agent is the best source of information about firearms regulations in your hunting destination, or you can consult that nation's embassy. If possible, carry copies of any regulations to show airline personnel.

✳ ***What to Pack:*** There's no more essential piece of luggage for the traveling hunter than an airline-approved, hard-sided, locking guncase. Even when traveling by road, carry your firearms in a hard case. The best sturdy cases should withstand abuse while keeping your rifle zeroed in, and should have room for other paraphernalia.

✳ ***At the Airport:*** The problem with guncases is that they look like guncases, and are ripe plums for thieves. "Take-down" shotgun cases look less so, and you can even fit a rifle into one by either having a take-down model or by removing a barreled action from its stock, if that won't disturb the bedding.

Upon arrival at the airport for check-in, declare your rifle and ammunition. Don't let a skycap tag your baggage without telling him you're transporting firearms. In fact, you'll probably have to check in at the counter. Show up well in advance of your flight, and have bolts out of rifles, shotguns broken down, and ammo (you're generally allowed ten pounds) packed in original boxes. Be prepared to demonstrate to an uncomprehending ticket clerk that your weapons are unloaded, and to explain the rules about airlines' accepting firearms and ammunition. You will be given a firearms tag to sign; slip this inside your guncase if possible. Have I.D. tags on all your baggage. Above all, do try to have a marvelous time.

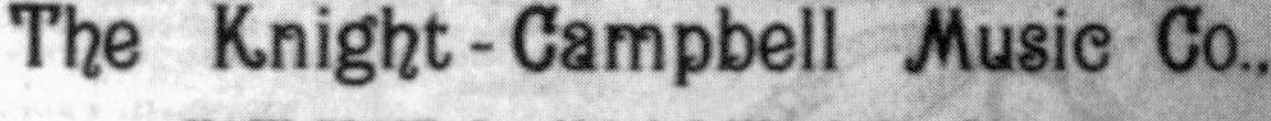

8

Preparation and Cookery

A venison stew, properly seasoned with the proper ingredients, is very satisfying and very filling. It does not lie on the stomach like a fry, nor need scraping like a broil, nor has it the smoky flavor of a roast by an open fire. It is taken in a semi-liquid form and percolates through the system from the palate to the diaphragm—or whatever you call it—filling the inner man with joy and gladness.

—R. Ritchie, *Sports Afield*, October 1898.

B asic Field Care

Like most manual procedures, the process of dressing a deer, while by no means a daunting one, is genuinely best taught through coaching—someone knowledgeable taking you through the physical task step by step, correcting you as you go, answering your questions on the spot as they arise.

The best way to learn how to dress a deer is to go hunting with someone who already knows and is willing to show you how. Most old-time hunters are more than happy to have someone along who is willing to dress-out their deer for them in exchange for the instruction. Otherwise, learning how to gut a deer by reading about it is a little like trying to learn how to ride a bicycle by taking a book from the library on the subject: It can be done, but eventually you have to get on that bike and begin pedaling.

There are certain basic objectives in the field care of game that ought to be kept in mind. For example:

✳ Remove the entire alimentary canal, from the throat to the anus, without letting any of its contents spill back into the body cavity.

✳ Keep the meat clean and make sure no loose hairs from the hide touch the meat. While not a serious problem at most times with deer, hair from a buck in the rut, if it gets on the meat, can impart the gamy taste for which venison is infamous.

How to Make an Indian Travois

When hunting remote sites, carry a small daypack with some heavy-duty twine, a small folding saw, and a couple of rags. With your hatchet or saw, cut two straight, slender saplings 10 to 12 feet long. Bind the two poles together at the small ends, then make a frame across the large ends, lashing two or three short pieces of wood crosswise to form the frame. Tie your field-dressed deer to the frame, place the extensions over your shoulders using the rags of padding, and start your trip out.

Travois Deer Drag

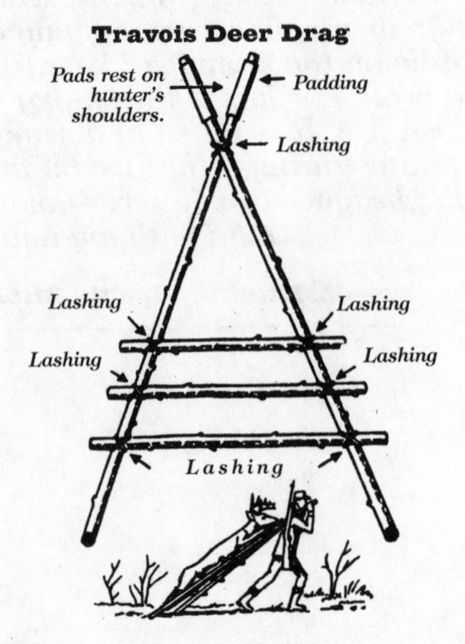

* Get a dry pellicle to form around the meat to prevent flies from laying their eggs in it.

* Get the carcass up off the ground so air can circulate around it and cool it down properly to prevent bacterial growth. A cold night with temperatures of around 40°F or lower is usually sufficient to chill a hanging dressed deer without your having to remove the hide. For larger animals such as elk, or in warmer weather, the hide must be removed and the meat gotten to a cooler as soon as possible.

* Let nothing of value go to waste.

Some Final Hints: If your deer is going to hang overnight, hang it high. There is nothing more disheartening than to discover the next morning that a coyote or a badger or a skunk has been at your deer and filled its belly on your meat.

Dusting the exposed meat with pepper is said to keep the flies away; but a tightly woven deer sack works well if you make sure to shake out any flies before tying it closed. You can make your own out of an old clean sheet from a double bed—sew it up like a giant pillowcase. It's both economical and reusable after washing.

When removing the lower hind leg from a deer—it's actually his foot, but it still looks like a leg—always try to hit the joint so you do not cut the back tendon. With this tendon gone, hanging the deer is difficult.

If you take your skinned deer to a butcher, try to bring the carcass in whole—it makes his job much easier. Chill it thoroughly, wrap it in blankets and sleeping bags, and it will keep cold for quite some time. A carcass should not be wrapped in plastic, as this retains heat and moisture.

Before delivering your deer to the butcher, take time to cut out all the bloodshot meat and tidy up the carcass as best you can. (A butcher may not be as finicky about the state of the venison you are going to eat as you are likely to be.)

If you are going to keep the cape for a trophy, always take enough hide for the mount. Cut around the body behind the shoulders and brisket and split

Drafty Deer Skinning

You can simplify the tedious task of skinning your deer by blowing air under the hide. Borrow a portable compressor with an air hose and a straight air nozzle (this blows a direct stream of air rather than fitting onto a tire valve). Make an incision in the deer hide just above the first joint on all four legs, insert the tip of the nozzle and inject air. The hide will inflate like a balloon and pop most of the subcutaneous adhesions that connect skin and meat. The hide then can be stripped free of the carcass with only a little more work than it would take to peel a banana.

the hide up the back of the neck. If you are unfamiliar with the arcana of "turning the ears" or "splitting the lips" on a cape, have your taxidermist show you the procedure on the next head he gets. In the field, salt will be needed to preserve the cape. A deer cape will require three to five pounds of table—not rock—salt rubbed into all parts of the skin side. Too much salt is better than too little. A cape can also be taken to the taxidermist frozen.

In the Crow tribe, after the meat was hung to dry, the women would make a tent from a buffalo skin by first soaking it for days in a solution of water and wood ash to slip the hair; then they would stake it to the ground and shampoo it with buffalo brains to tan it. They would let it dry, then use a sharpened buffalo scapula to grain and thin the hide. Lastly they would smoke it to keep it supple. When the sun fell on it, light would glow through its translucence to brighten the interior of the lodge, something of value retained and a commitment fulfilled.

Better Skinning

This is a simplified method of removing the hide from a deer. All you need is a rope, a secure place to hang the deer, a stick or gambrel to hold the hind legs apart, a saw and a sharp knife.

✳ ***Step 1:*** Before hanging the deer, cut through the skin all the way around the rear legs just below the joints. Then cut along the inside of the thighs from those circular cuts up to the abdominal cavity. Pull the hide from the legs, using your knife when needed.

✳ ***Step 2:*** Remove the rear legs with a saw below the joints. Insert a stick or the hooks of a gambrel into the joints between the bone and the tendon; then hang

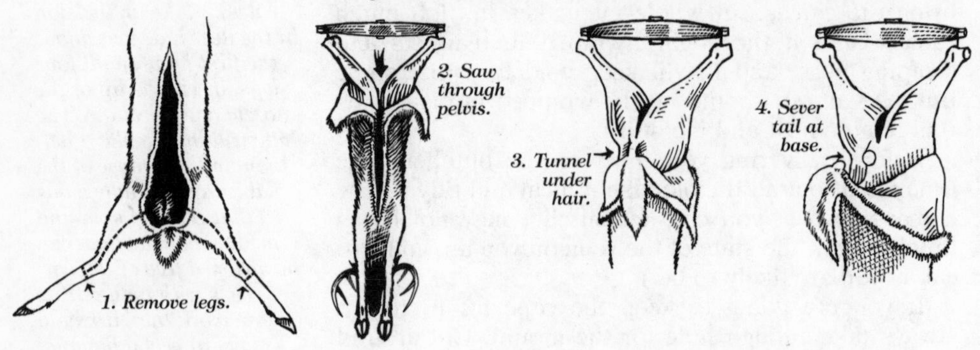
1. Remove legs.
2. Saw through pelvis.
3. Tunnel under hair.
4. Sever tail at base.

the deer upside down. (If you don't have a friend to help you, use a small pulley to hoist the deer.) Use the saw to cut down through the aitch bone at the pelvis.

✳ ***Step 3:*** Continue skinning. When you reach the tail, do not cut it off. Instead, skin around the tail and use your knife to "tunnel" under it.

✳ ***Step 4:*** Partially skin out the tail, then cut it off at the base. This technique ensures that loose hair will not stick to the carcass.

✳ ***Step 5:*** Grab the hide in one hand and pull down, simultaneously jabbing the carcass at the point where the hide hangs off.

✳ ***Step 6:*** Saw down from the abdominal cavity through the breastbone. Cut through the skin around front legs above the joints, then cut along inside of legs to the breast opening. Skin out legs.

✳ ***Step 7:*** Saw off the front legs above the joints. Continue to pull the hide down, using your knife to loosen the skin from the neck. Skin down to the base of the skull and saw off the head.

✳ ***Mounting:*** If you intend to mount the deer, cut along the top of the neck as shown instead of through the breast. Remove the head, with hide attached, and deliver to the taxidermist.

Hairless Venison

When skinning a deer, hunters find it almost impossible to keep hair off the carcass. Even wiping the meat with vinegar and water doesn't always remove the hair, especially in the crevices and body cavity. An easy method for removing stubborn hair is to use a propane torch. Don't cook the meat—just use the flame to singe the hairs.

A Bicycle Rack for a Buck

Many hunters go off deer hunting with not even a thought of how they will carry their trophy home if they connect. With today's automobile styles, tying down a deer for the long trip home can be perplexing, since there are few places to lash onto. If you don't do it right, it could mean the trophy falling off at 55 mph or some expensive scratches in the paint on your automobile.

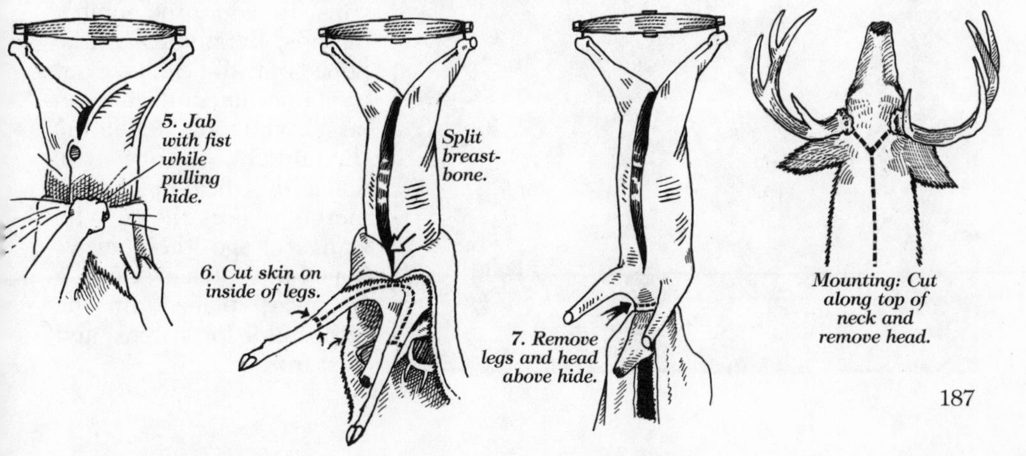
5. Jab with fist while pulling hide.
Split breast-bone.
6. Cut skin on inside of legs.
7. Remove legs and head above hide.
Mounting: Cut along top of neck and remove head.

Next time you go deer hunting, bolt a versatile bicycle carrier to your bumper. A deer can be easily lashed to this in a number of ways; and it can be carried home safely without problems.

One method is to attach the rack to the back bumper and secure the safety strap to the trunk lid of the car. The deer can be laid on its back between the rack and the trunk of the car. This takes some of the stress off the bike rack and distributes it to the vehicle. Then legs and head or antlers can be lashed to the upward bicycle frame holder of the rack so they don't scratch the car. Before you transport, check the bumper attachments and lashing for tightness.

Four Better Ways to Pack Out Game

In short, there are several good reasons to get game out of the woods by methods least injurious to the meat, to oneself and to the environment. The choices fall into four groups, all of which employ mechanical assistance: packing, dragging, barrows, and mountain bikes.

✳ ***Packing:*** For the backcountry hunter, this method requires no extra equipment. You pack in food, clothing and camping gear with a large frame pack; you pack out meat in the same pack. The day hunter, on the other hand, can return to the roadhead, exchange rifle or bow for pack, and hike back to the kill. The animal's quarters are now placed either in the pack bag itself, or tied directly to the frame. The only tools necessary are a knife, a small saw and enough light cord to secure the quarters. With an average-size backpack, most deer can be carried out in two or three trips. To reduce the number of trips, the animal can be boned out. You can take out a jumbo mule deer or a large whitetail, including the antlers, in one trip, using this field-butchering method. For those who want to age their meat, boned-out steaks can be wrapped, placed in the refrigerator for a week, and then frozen.

Dragging a deer with his head up helps keep antlers from breaking.

Deer transport shouldn't be a drag.

✳ ***Dragging:*** After the first snowfall, the easiest way to transport meat is with a sled. Choose a lightweight, plastic toboggan without runners, the kind designed for kids and that costs about $10 in most variety stores. They are virtually indestructible and can be pulled over rocks without fear. Drill some holes on the edges for lashing down the load.

✳ ***Barrows:*** Called "game carriers," these modified wheelbarrows are made from aircraft-grade aluminum and can accommodate animals up to 500 pounds. They are helpful to the hunter who doesn't like quartering animals in the field, and can be useful in moderate-to-hilly terrain, and on trails. In really steep country, they're hard to handle, especially if fully loaded.

✳ ***Mountain Bikes:*** Permitted only in country that isn't designated wilderness, a mountain bike lets the hunter reach his destination in one-quarter the time it takes to walk there. Used in conjunction with a frame pack carried on his back and panniers attached to the bike, a hunter can transport meat with little effort, even up and down very steep grades, since mountain bikes have 21 gears and superb brakes.

Don't Go for the Jugular

Those who maintain that bleeding game improves the meat should know that merely cutting a deer's throat doesn't help much, since the veins and arteries there lead only to the neck and head. Most of the blood is at the other end of the animal.

The blood vessels that run to the back and hind quarters are connected to the top of the heart. They make a U-turn before going very far into the neck, and slitting the animal's throat will not sever them.

The way to bleed an animal is to stick it: Insert a knife blade, point first, into the base of the neck, from the underside, at a slight inward angle, toward the heart. After penetration, turn the blade and cut an inch or two in each direction until the aorta is severed. You can't see this artery, but you'll get a rush of blood when you hit it, so stand back. Hanging the animal by the hind legs or putting its head down on slop-

Venison on Ice

Successful early season archery hunters are often faced with the dilemma of warm temperatures and possible meat loss due to bacterial contamination. Carefully placing a 20-pound bag of ice in the chest cavity and another in the abdominal cavity of the deer will eliminate any chance of spoilage. The cold generated by the ice will soon migrate through the muscle tissue, ensuring safe arrival home. Do not skin the animal. The hide is an efficient insulator that will hold the lower temperature in the body cavity. If you hunt remote areas, you can store ice in a cooler prior to the hunt.

ing terrain will drain the blood. Any good field-dressing knife will do. A drop-point with a sharp four-inch blade is perfect.

This is not an argument for bleeding. Many people believe that you should begin field-dressing as soon as possible instead of wasting time with bleeding an animal. But if you prefer to bleed your meat for one reason or another, do it properly instead of just cutting the throat.

European Skull Mount

If you've bagged a deer with a head not quite worthy of full taxidermy treatment, why not try a European skull mount? You can do the work yourself—with no special equipment. All extraneous parts, including hide, hair, brains and ligaments, are removed. The skull should then be boiled to loosen remaining tissue and bleach the bones. While boiling the skull, hold the antlers out of the boiling water to avoid discoloration. An outside cooker is ideal. The mount can be displayed immediately.

Pack It All Out

For some reason, Eastern hunters usually drag deer carcasses, even across long distances. A better technique is to pack out just the deboned meat. A light aluminum packframe equipped with a nylon rucksack measuring 24 x 18 x 12 inches will hold all the meat from a large mule deer or whitetail buck.

The next time you have a deer professionally butchered, ask the meat cutter if you may watch. Most will be glad to have the audience. Or you can purchase a book or video outlining the necessary procedures for butchering big game.

Start by skinning the side that faces up. Lay the skin out straight from the spine. This will act as a buffer from dirt contamination. Line the pack with a muslin game bag, and as you debone the meat, put it into the pack. Work from the head to the rump. When the first side is finished, roll the deer over and repeat the process.

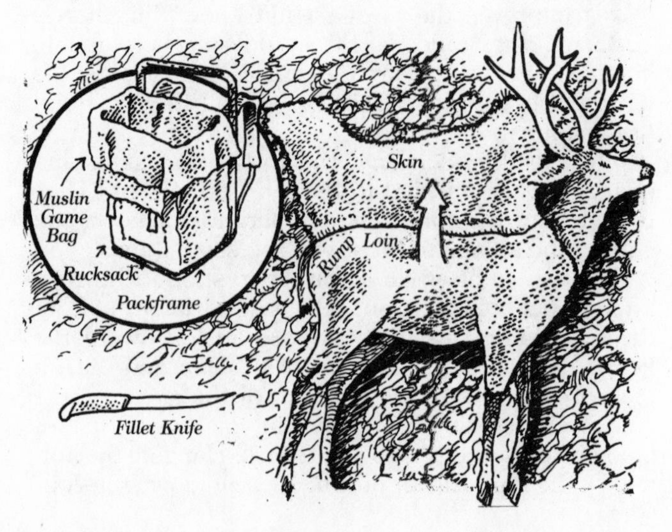

A long, thin blade, such as a filleting knife, works best because you can work closely around the bones. The rump, and the loin along the spine, should be removed in one piece. Shoulder and neck pieces can be used for ground meat and stew. Once the meat is back to camp, separate and cool quickly.

Deer Transport Tips

At its most basic, the intended outcome of hunting is the permanent interruption of an animal's opposition to the force of gravity. In short, drop him.

To drag a deer out by its front legs, run a sturdy, sharpened stick between the tendons.

Once an animal, such as a deer, is on the ground, though, the responsibility for his transportation falls to the hunter who put him there. It would certainly simplify matters if all deer fell where you could reach them with a pickup truck, or if we all had access to horses to carry our burdens for us. But deer seem to have this almost wanton disregard for our fondest hopes when we go out to hunt them, so we need to be prepared for the very real possibility of having to use our own legs and backs when it comes to transporting dead game out of the field. By thinking things through, however, we can bring down to entirely manageable levels the task of moving almost any deer:

✳ Upon reaching the deer, first make certain he is dead, then tag him properly.

✳ If you are going to drag a deer any distance, the pulling will be easier if you first field-dress him and remove his front legs at the knees and hind legs at the joint just below the hock. (Don't detach the Achilles tendons from the hind shanks; you'll need these for hanging the deer or carrying the hindquarters.)

✳ Survey the area carefully before deciding which way you are going to drag or carry your deer: Hunters sometimes wear themselves out hauling an animal uphill when, had they just looked around, they could have easily dragged him down to a road or gate.

✳ If you have to quarter a deer to pack him out, try to leave his hide on, to keep the meat clean, and cut him into pieces as large as you can handle.

✳ Don't use plastic trash bags for carrying meat:

Sign Your Game Tag Immediately

Most states require that the hunter's game tag be removed, dated and signed immediately after an animal is harvested. What to do if you're without a means of signing the tag? First, try writing with the lead tip of a bullet. If this fails, try dipping a weed or stick in the animal's blood.

The meat can sour in plastic, and trash bags are often treated with chemicals.

✱ Carry antlers with the points down, both for ease of movement and for safety.

✱ Also, flag antlers with orange tape or cloth.

From Field to Oven: Five Steps

Skinning: Remove the skin, especially on large animals like elk and deer. Doing so prevents the meat from spoiling due to retained body heat. Try to field-dress and skin out immediately, usually within 30 minutes of the kill.

✱ ***Washing:*** Hair, body wastes and dirt can affect the taste of surface cuts of meat. Bring a clean plastic tarp on your hunts. Lay the carcass on the tarp and wash the meat thoroughly as soon as possible. On hunting trips away from home, use a hose in a rancher's yard (with permission, of course).

✱ ***Quartering:*** Reducing a carcass to individual quarters makes hauling, washing and hanging much easier. Cutting an animal into five sections is the most practical approach: two hindquarters, two front quarters and the rib cage. It is important to cover each quarter with a muslin gamebag.

✱ ***Hanging:*** Hang elk and deer for 12 days in a commercial meat locker, where controlled temperatures allow the enzymes in the meat to break down the structure and tenderize it. (Recent studies have shown that antelope is less affected by hanging than elk and deer.)

✱ ***Cooking:*** Because venison, elk and antelope are dry meats, their flavor may be enhanced with a good wine sauce or a conventional brown gravy. However, you won't avoid the gamy taste by burying wild meat in a fancy sauce. The single best way to guarantee a delicious game meal (after proper preparation) is to avoid overcooking the meat—keep it pink. Generally, when wild game goes completely gray, flavor and tenderness go too!

No-Sweat Drag Harness

Hauling out a full-grown animal alone can take the strength and endurance of an Olympic athlete, but you'll find the job easier if you use more of your large leg muscles and less of your arms and shoulders. All you need is a 10-foot length of rope and a heavy flannel shirt. With these you can make a hip belt to allow hands-free dragging. Just roll the shirt (it must be a heavy one) into a four- to six-inch band you can tie around your waist. Attach the rope to this hip belt and tie yourself to the animal. You'll be amazed at how much easier the drag duty is.

Keep the Gaminess out of Your Game

Your frozen venison seems to taste more gamy the longer it's in storage? Unfortunately, animal fat and bone marrow will not freeze completely at temperatures found in the typical home freezer. Unless you can store game at subzero temperatures, slow decomposition takes place and spoils the flavor.

What to do? Certainly it helps to date your packages to be sure those subject to spoilage are used within a few months. With venison (or other big game) remove all bones and fat before wrapping for freezing. Finally, use up any game before the new season—even the best deep-freeze won't keep meat forever, and a full year is about the maximum.

Perfectly Aged Venison

Venison can be aged to improve the quality of the meat. Aging carcasses for seven to nine days at 34°F to 37°F improves tenderness and changes flavor. It's much better than the 24-hour chill, cut and freeze method.

You should consider these factors in your decision to age or not to age:

✳ ***Age of Animal:*** Young animals are fairly tender, so aging is not necessary. Aging older animals improves tenderness.

✳ ***Stress Before Death:*** If the animal has run a long distance before death, the meat may have a high pH, and the potential exists for increased bacterial growth during aging.

✳ ***Aging Facility:*** A clean facility where temperature can be controlled is essential. Freezing temperatures stop aging. Aging temperatures above 40 degrees result in spoilage.

✳ ***Intended Use:*** If all of the meat is to be ground

Using Your Deer

Once trophy considerations have been addressed, a deer hunter usually intends to eat or share all the salvageable meat, including the tongue, liver and heart. But what becomes of the remainder? Too frequently, many deer parts are discarded. Those who waste that which can be used lose a very substantial and wonderful part of hunting. Here are a number of ideas for those who want a bit more out of our sport. The deer's antlers, skin, tail, glands, bones and virtually every other part have many uses. If you glean only one idea from this, or are stimulated to develop others, we're all better for it.

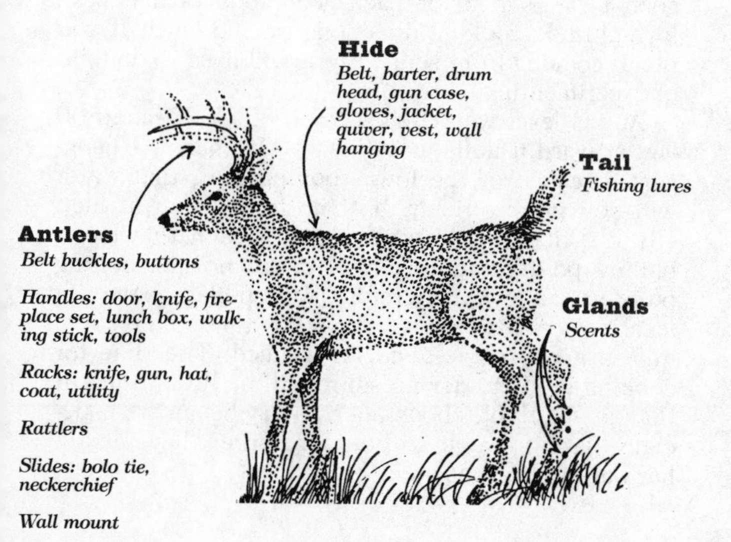

into sausage, aging is not necessary. Consider aging only if you want steaks and roasts.

Venison in the Diet

Venison has less fat than farm-bred meats and is an excellent source of protein and iron. For a comparison, here are the nutritional breakdowns for 3 ½-ounce servings of venison, beef and chicken.

✱ ***Calorie content:***

Venison, 158; beef, 222; chicken, 190.

✱ ***Protein:*** 67 percent of the daily requirements for both venison and beef, 64 percent for chicken.

✱ ***Fat:*** Venison, 18 percent; beef, 42 percent; chicken, 35 percent.

✱ ***Cholesterol:*** Venison, 112 milligrams; beef, 90 milligrams; chicken, 89 milligrams.

✱ ***Iron:*** 25 percent of the daily requirements for venison, 18 percent for beef, 7 percent for chicken.

Venison's low fat and calorie content more than compensate for its slightly higher cholesterol.

High-Altitude Cooking

Most recipes are designed, and cooking time given, for use at or near sea level. Most need no modification up to 2500 to 3000 feet, but higher altitudes can cause many problems. It is often necessary to adjust the amount of baking powder and flour. Because such adjustments vary from recipe to recipe no set guidelines are possible. Therefore, it's a good idea to look for packaged products that have high-altitude cooking directions on the label. If you prefer cooking from scratch, you will need to do a bit of experimenting.

At sea level water boils at 212°F. With each 500 feet upward it boils at 1°F less. This seems to suggest, because of the lower boiling pint, that water will boil more quickly, but the fact is that at high altitudes it takes water longer to reach this lower boiling point than to come to the normal boiling point at sea level. Therefore, food boiled above sea level will require longer cooking time than usual unless a pressure saucepan is used. The time for cooking foods under pressure will be little different than at sea level. Foods such as dry beans are especially hard to cook without pressure; they remain hard after hours of boiling.

SPORTS AFIELD

THE VERMONT DEER SEASON.—CAMP CHORES.

SPORTS AFIELD PUB. CO.,
358 DEARBORN ST., CHICAGO.

Cookery

Deer Camp Chili

One of the best meals you can serve in a hunting camp is venison chili. It should be cooked on a woodstove the evening before and served the next evening with hard-crusted sourdough bread, along with an appropriate brew. The recipe will serve eight hunters, and can be halved for an average family.

---

1 lb. dried pinto beans
2 lbs. (about 4 cups) venison, cut into ¾-inch cubes
3 or 4 tbsps. bacon fat or cooking oil
2 cloves garlic, finely chopped
4 large or 6 medium onions, coarsely chopped
2 tsps. ground cumin
1 or 2 tbsps. hot chili powder, Mexican-style
1 28-oz. can crushed tomatoes
1 28-oz. can peeled tomatoes, cut into chunks

Wash the beans and place them in a pot with 1 ½ teaspoons of salt and enough water to cover them well. Bring to a boil. Remove from heat and soak for about an hour. Add more water, if necessary, to just cover the beans, and simmer while preparing the rest of the chili (about 1 hour).

Salt the venison cubes in a large, heavy iron pot, brown them with 3 or 4 tablespoons of bacon fat or cooking oil. Add the garlic, onions and spices. Cook covered, until the onions and garlic are soft. Add both cans of tomatoes and heat until simmering. Then add beans, including the bean juice (two cans of cooked pinto beans may be substituted). Bring to a slow boil, then lower the heat and simmer for 3 to 4 hours. (It's even better if served reheated the next day.)

Ladle into serving bowls and sprinkle with chopped raw onions or coarsely shredded sharp cheddar cheese. Serve with crusty fresh-baked bread or hot cornbread and butter.

Tender Venison Steaks

1 lb. venison steaks, about 1 inch thick
1 cup all-purpose flour
1 tsp. salt, 1 tsp. pepper

MARINADE
1 tsp. Worcestershire or steak sauce
2 tsps. salt
¾ cup red cooking wine (12 percent alcohol)
1 ½ qts. water

Marinate the steaks overnight in the above mixture, rinse thoroughly in clear water, and drain.

In a clean paper lunch bag mix together and shake 1 teaspoon salt, 1 teaspoon pepper and 1 cup all-purpose flour. Add steaks one at a time and shake bag to coat each one thoroughly. Heat some cooking oil in a frying pan. Fry the steaks over medium-high heat for 10 to 12 minutes on each side. Drain them on paper towels before serving.

For gravy, keep the pan drippings hot. Add 2 tablespoons margarine and let it melt. Add ¼ cup all-purpose flour and stir immediately with a large spoon to break up lumps. When mixture is smooth (a couple of minutes), add 8 ounces hot water and stir until thickened to your liking.

Venison Purloo

Whether it's called pilaf, pilau or purloo, meat cooked with rice is a favorite the world over, and an ideal way to tenderize tough cuts. Try the recipe below whenever you bag a deer.

1 deer heart
1 set deer kidneys
5 cups water
1 medium onion, chopped
1 tsp. salt
½ to 1 tsp. red pepper flakes
1 ⅓ cups long-grain rice

Trim and chop the heart and kidneys into ¾-inch cubes. Put the meat into a suitable pot and add the water, onion, salt and red pepper flakes. Bring to a boil, reduce the heat, cover tightly, and simmer for about 1 ½ hours, or until the deer heart and kidneys are tender. Add the rice, bring to a new boil, reduce the heat, cover, and then simmer for 20 minutes. Remove the lid and continue simmering until all the water is absorbed into the pilaf. (If you have to eat with chopsticks, use short-grain rice, which is a little stickier.) Serves 4 to 6.

Deer liver can also be used, but it won't require long simmering. Add it after the heart has simmered for 1 hour or so.

Watch Your Tongue

There are several good ways to cook deer tongue. Here's an easy recipe for beginners. When dressing the deer, cut out the tongue and leave the skin on it. (You should always take the tongue, as well as the tenderloin, at the end of the field-dressing operation.) Boil the tongue (or tongues) in 2 quarts of water to which you have added a chopped onion, 3 bay leaves, 3 cloves, 1 tablespoon salt and ¼ tablespoon of red pepper flakes. If you have a tongue from a large deer, boil it for 1 ½ hours; from a small deer, 1 hour, or until it is tender. Let the tongue drain. Skin and slice it crosswise, into wheels about ¼-inch thick. Serve the slices on fresh crackers, topped with a little Creole seasoning or brown mustard.

Delicious Venison Paté

Serve this delicious venison spread at your next party, and everyone will demand the recipe. It's an excellent way to use a leftover roast.

1 lb. well-done venison, cold, shredded in a blender or meat grinder
1 8-oz. container French onion dip
2 tbsps. Worcestershire sauce
1 tsp. garlic powder
1 tsp. parsley
½ tsp. black pepper

Blend all the ingredients thoroughly, chill for a minimum of 2 hours, then serve on party crackers. This also makes an excellent sandwich spread.

A Simple Marinade

Many venison recipes call for elaborate marinades, but one of the simplest works as well as any. A mixture of lemon juice and olive oil both tenderizes and adds needed fat to the venison.

Many cooks have used this marinade on roasts, steaks and stew meat. It works on every cut of venison and tenderizes without adding a heavy lemon taste.

For 1 pound of stew meat, use about ⅛ cup of lemon juice, mixed with ¼ cup of olive oil. Leave venison in the marinade for 3 to 4 hours in the refrigerator. It can then be removed and cooked, using whatever venison recipe you like best.

Homemade Barbecue Sauce

2 cups ketchup
2 tablespoons hot sauce
¼ cup Worcestershire sauce
1 tbsp. garlic powder
½ cup brown sugar
Stir mixture well and bring to a boil for about 2 minutes. Let it cool and thicken. This sauce adds flavor to any barbecued meat.

Marinated Ribs

Marinate one rack of deer ribs overnight in a mixture of ½ cup red cooking wine, 1 tablespoon salt, and enough cold water to cover. The next day, rinse well in clear water. Boil the ribs for 1 hour, adding to the water 1 teaspoon each of salt, pepper and sugar.

Mix the sauce while the ribs are boiling: 1 can of beer, ½ cup honey, 1 teaspoon salt, ½ teaspoon pepper, 2 teaspoons lemon juice, and 1 cinnamon stick (to be removed after sauce is cooked). Cook sauce just long enough for ingredients to be well blended.

When ribs are done, remove them and let them cool. Reserve ½ cup of the sauce. Pour the remaining sauce over the ribs and marinate them for 1 hour (do not refrigerate). Add 1 teaspoon each of ketchup and mustard to the reserved sauce, mix well and pour over the ribs. Bake at 350°F for 30 minutes, or until tender and brown.

Note: Do not overcook. Also, do not let time elapsed between boiling and baking the ribs exceed 2 hours. Do all this right and you will be in for a treat.

Booker Noe's Venison With Bourbon

Booker Noe of Bardstown, Kentucky, has hunted and fished and camped and cooked for over half a century. He is heir to the famed Jim Beam distilleries started six generations back, in 1795, by Jacob Beam, who followed Daniel Boone's trail over the Cumberland Gap to homestead in that fertile wilderness. Jacob brought with him an expertise for turning the local corn, rye and barley into a spirit he named for the area: Bourbon County. The rest, as they say, is history. Booker recently introduced an uncut, unfiltered sipping bourbon, drawn straight from the barrel. It is called, simply, Booker's. You might use a less expensive bourbon for his venison dish.

1 leg of venison
salt and pepper for rubbing
1 cup flour
1 stick butter, melted
¼ cup bourbon
¼ tsp. garlic powder
¼ tsp. black pepper

Rub the leg of venison with salt and pepper. Mix the flour, butter, bourbon, garlic powder and black pepper to make a paste. Place the venison on a rack in a broiling pan and cover thoroughly with the paste. Refrigerate until paste becomes firm. Roast at 350°F, covered.

Venison Heart and Liver

Too many deer hunters who enjoy the wild tang of venison are concerned only with tasty chops, steaks and roasts. They leave the hearts and livers to coyotes and ravens. If properly prepared, though, these organs are very tasty.

Place the whole liver in your freezer just long enough to give it firmness. Then with a sharp knife slice at a slight angle, making sure the slices are reasonably thin. Now here's the culinary secret: Before cooking, place slices in boiling water for about 15 seconds. Remove and dry on paper toweling.

Roll liver in slightly seasoned flour and place in skillet with heated cooking oil or (preferably) bacon grease. Fry for approximately 4 minutes and serve immediately (with fried potatoes, coarse wheat bread and coffee). If you prefer that extra touch, garnish the platter with bacon strips and onions.

Now for the heart. First place it in a mild solution

of salt water for a couple of hours to draw out excess blood collected in the cavities. Remove, rinse and place it in a medium-sized kettle of fresh water. Add 1 bay leaf, 1 teaspoon allspice, 1 small diced onion, 1 teaspoon salt, a diced carrot and ½ cup diced celery and leaves. Cover and simmer for 2 or 3 hours, depending on the size of the heart.

After it is thoroughly cooked, let the heart remain in the broth until it's cool. Then remove and place in the refrigerator to chill. Slice thinly, and serve as hors d'oeuvres on bread or crackers, garnished to your taste.

Venison Sausage Quiche

Here's a venison recipe for any occasion, but it is especially recommended if you're introducing people to venison. They'll be amazed at how it livens up the quiche. Here's what you'll need:

butter for frying
1 lb. fresh mushrooms, sliced
¼ cup onions, chopped
1 lb. venison sausage mix (see below)
4 medium eggs
2 9-inch pie shells
½ pound grated Monterey Jack cheese
1 cup heavy whipping cream
salt and pepper to taste

Preheat the oven to 350°F. Melt a little butter in a frying pan and sauté the mushrooms and onions for a few minutes, then drain and set aside.

Brown the sausage and set aside. Beat the eggs slightly in a bowl and set aside.

Bake the pie shells at 350°F for about 8 minutes, or until they are golden brown. (If you buy ready-to-use shells, follow the directions on the package.) Mix the sausage, mushrooms, onions, cheese, eggs and cream in a bowl. Add a little salt and pepper to taste. Then spread this mixture evenly into the two pie shells, and bake them at 350°F for about 30 minutes. The pies can then be cooled (it takes about 1 hour), wrapped in foil, and frozen. Bring them out for breakfast, lunch or dinner. Serves 5 or 6.

Venison Sausage Mix

3 lbs. ground venison
½ lb. ground pork
1 tbsp. sage
1 tbsp. thyme
1 tbsp. black pepper
1 tbsp. marjoram
½ tbsp. salt

Combine all ingredients. Refrigerate or freeze the mix until ready to cook.

How to Make a Deer-Neck Stew

The bony neck of a whitetail or mule deer does contain some of the best meat on the animal. The neck is usually best for stew, since the bone and its marrow add wonderful flavoring. You can use the neck whole, if you've got a pot of suitable size. If not, the neck should be cut. This is easy with a meat saw or a hacksaw with an uncoated blade. Or take your animal to a meat processor. Freeze it in 5-pound packages.

4 or 5 lbs. of deer neck
1 beef bouillon cube
3 medium onions, quartered
3 medium potatoes, quartered
3 carrots, sliced
1 stalk celery, sliced
1 tbsp. parsley, chopped
salt and pepper to taste

Heat enough water in a dutch oven to cover the neck bones and dissolve the bouillon cube in it. Bring it to a boil, reduce the heat, cover and simmer for 2 hours. While the meat is cooking, peel and quarter the onions and potatoes. Add all the vegetables and the parsley to the dutch oven. Bring it to a new boil, reduce the heat, cover and simmer for an additional ½ hour. Remove the neck pieces and pull off the meat with a fork. Discard the bones and put the meat back into the dutch oven. Stir in a little salt and pepper and simmer for 15 more minutes.

Leftover Venison for Breakfast

Didn't quite finish that venison roast or steak? Refrigerate the leftover meat, and in the morning cut it into bite-sized pieces. Chop some onion (if you have small green onions, use about half of the tops). Melt a little butter in a frying pan and sauté the onion until it is tender. Add the diced venison and stir for a few minutes. Add a little hot black coffee and bring to a boil. Reduce heat and add salt and pepper. Then, sprinkle in flour very, very slowly, stirring constantly, until you get gravy the way you like it. Add more coffee if necessary. Serve over hot biscuits.

This stew makes a complete meal when served with sourdough bread.

The Venison Omelet

This great omelet combines the taste of venison, sharp cheese and onions. Ingredients listed are for one omelet.

---

2 eggs
1 tbsp. milk
dash of soy sauce
¼ cup venison scraps, chopped
1 small onion, chopped
¼ cup feta or other sharp cheese
cooking oil or butter

Break eggs into a bowl and beat with a fork. Add the milk and soy sauce. Finely chop the venison and onion and slice the cheese. Heat cooking oil or butter in a frying pan; when thoroughly heated add the egg mixture. As the eggs begin to set, place the venison, onions and cheese on half of the omelet and fold the other half over. The heat may be reduced after the omelet is folded. The omelet is ready when the filling has warmed and the cheese has melted.

Camp Cooking

If your hunting cabin or fishing camp has electric power, you should have an electric cooking pot. They cook at very low heat for long periods of time. Thus you can easily put a venison roast, frozen or not, into the pot early in the morning, go hunting or fishing all day, and have a very good meal ready at night.

This method of preparation is an excellent way to cook any meat thought to be tough. Cooking the meat all day under a tight cover at a low heat definitely beats zapping it for a few minutes in a microwave. Almost any pot-roast recipe can be used. Or you might want to try this one:

5-lb. roast—deer, elk, moose, etc.
salt
1 envelope onion soup mix
¼ cup water

Before you leave camp in the morning, pour the ¼ cup of water into the pot and set the pot on low. Place the roast in the pot and sprinkle it with a little salt and the soup mix. Put the cover on the pot and leave the pot on until you get back that night. If you prefer, add potatoes, onions, tomatoes and such—fresh, frozen or canned. Fresh or frozen vegetables should be cooked all day, but canned goods can be added at sundown or after.

Venison Stroganoff

2 lbs. venison, cubed
flour
2 tbsps. bacon grease
1 envelope beef onion soup mix
2 ½ cups very hot water
1 can cream of mushroom soup
3 tbsps. sour cream
egg noodles

Remove all the fat from the venison and cut the meat into bite-sized cubes. Roll the cubes in flour to coat lightly. In a large iron skillet over medium heat, melt the bacon grease and brown the venison on all sides.

Mix the beef onion soup mix with 2 ½ cups of very hot water and pour over the venison. Cover and simmer for 1 ½ to 2 hours, stirring occasionally and adding water if necessary.

When the venison has cooked, add the mushroom soup and sour cream. If you prefer, add water for a thinner sauce. Cover and simmer while you prepare the noodles.

Cook the egg noodles according to

the package directions. Drain and place them on a serving platter. Ladle the Stroganoff over the noodles and serve immediately. Serves 4.

The Ultimate Venison Burger

Most deer hunters have a favorite recipe for adding a different meat to ground venison. Many use pork, or pork and veal. This is okay, but most people tend to overcook it because of a fear of contracting a parasitic infection from the pork. Overcooking, however, negates any benefit that adding the different meat may have had on the finished hamburger.

Instead, use inexpensive (but choice grade) beef suet to the venison during the grinding process. A ratio of 1 pound of suet to 5 pounds of venison will ensure juicy hamburgers. You can purchase beef suet at any supermarket or butcher shop; just tell the butcher the suet is for human consumption.

Process the venison and suet through the grinder once, then mix it thoroughly by hand. Grind it a second time. This second grind is very important and is often overlooked by amateur meat cutters.

Cook this ground-meat combination as if it were beef, using it in meatloaf and hamburgers.

Swiss Venison

The term "Swiss steak" refers to an inferior cut of beef that is at its culinary best after long simmering. Since venison is often a little tougher than modern feedlot beef, any steak cut from the leg of a deer can be cooked to advantage by a good recipe for Swiss steak. Allow from ⅓ to ½ pound of meat per person.

2 lbs. venison steaks
flour
1 stick margarine or ½ cup cooking oil
2 lbs. onions, chopped
2 large tomatoes, chopped
1 stalk celery, chopped
2 tbsps. parsley, chopped
¼ cup Worcestershire sauce
salt and pepper to taste

Cut the steaks 1 inch thick and pound them with a meat mallet. Then dust with flour, salt and pepper.

Melt 1 stick of margarine (or use ½ cup of cooking oil) in a skillet. Brown steaks lightly on both sides, then transfer them to a heated platter.

Brown the chopped onions in the skillet, then add the steaks, topped by tomatoes, celery, parsley and Worcestershire sauce.

Reduce the heat, cover and simmer for 1 ½ hours.

Serve the steaks with mashed potatoes and gravy from the skillet, along with fresh vegetables, salad and bread of your choice. If the gravy isn't quite the consistency you like, thicken it by stirring in a little flour mixed with water, or thin it by adding plain water or coffee. Serves 4.

Low-Salt Venison Jerky

Jerky recipes abound, but most contain too much salt. This delicious version is relatively salt-free. It requires 5 pounds of meat.

5 lbs. boneless venison or beef

MARINADE
¼ cup soy sauce
¼ cup water
2 tbsps. Worcestershire sauce
½ tsp. black pepper
½ tsp. garlic powder
1 tsp. onion powder
1 tsp. hickory smoke-flavored salt
(For less salt, substitute liquid smoke.)

Trim all fat and connective tissue from boneless venison (or beef). Cut the meat into ⅛- to ¼-inch slices. Cut with the grain for the traditional tough jerky, across the grain for a softer, less chewy product. Use pieces of meat scraps, too, if they're not going to be ground or used for stews.

Place the meat slices and marinade in a leakproof plastic bag. Force the air out of the bag and seal it tightly. Knead and tumble the bag gently to mix the sauce and meat, then place it in the refrigerator overnight. Remove and separate the meat slices and place them flat on the racks of a food dryer. Dry until the jerky is of the desired consistency. This will vary with the size and thickness of the meat slices. The jerky can be stored for a year or more in a covered—but not airtight—container.

Smoked Venison

A Utah woodsman and wild-game chef developed this technique for preparing venison, which ought to be in every deer hunter's recipe book.

---

electric smoker
5 to 10 lbs. lean venison meat
1 cup salt
½ cup brown sugar
molasses
vinegar
cheesecloth
black pepper to taste
wood chips (hickory, hard maple, apple, cherry)

Cut the meat into strips 6 to 8 inches long and no more than ½ inch thick and 1 to 2 inches wide. Soak the strips in saltwater-and-vinegar brine overnight in a deep bowl or crock.

Mix 1 cup salt and ½ cup brown sugar in a flat pan. A cake pan or cookie sheet works well. Dredge the meat in salt and sugar, rubbing them well into the meat. Next, rub ½ teaspoon of molasses onto each meat strip, then roll the entire quantity of prepared meat in cheesecloth. Place in a cool place for 4 to 5 hours.

Rinse the excess salt off the meat with cold water after the previous time allowance. (If all the salt is not rinsed off, the meat will be too salty to eat.) Dry each piece of meat with paper towels. Next, sprinkle black pepper sparingly on the strips. Your meat is now ready for the smoker. (Rubbing a small amount of molasses and brown sugar into the meat again will give it a sweeter taste after smoking.)

✳ Place the meat on racks in the smoker and smoke for 5 to 8 hours, depending on taste. Check the wood chips every hour to keep the meat under a constant smoke.

Venison Ragout With Cream Sauce

6 lbs. deer neck cut in 2-inch slices, with bones
5 oz. bacon grease (or oil)
4 medium-size onions, chopped
4 cloves garlic, finely chopped
6 oz. tomato paste
1 cup red wine
1 cup beef stock (from concentrate)
1 lb. fresh mushrooms, sliced
1 tbsp. vinegar
1 pinch sugar
1 tbsp. basil, dried
2 cups sour cream
salt, ground pepper to taste

MARINADE
approximately 5 cups buttermilk
1 tsp. juniper berries, crushed
5 black peppercorns, crushed
1 bay leaf
2 tbsps. fresh lemon juice

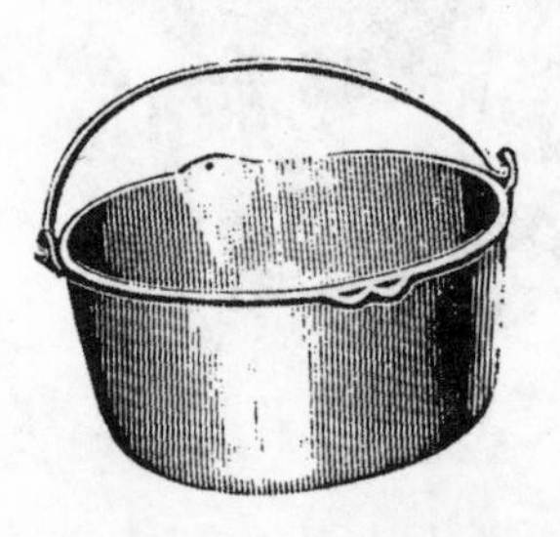

Wash venison parts with cold water, remove all bone splinters, then wipe the meat dry.

Mix marinade ingredients in a deep bowl. Submerge the venison parts. Cover the bowl slightly (not airtight) and store in a cool room for two days, or in a refrigerator for three days, flipping the parts daily.

After marinating, drain and discard buttermilk. Wipe the neck parts dry.

In a wide pot, brown the venison parts in bacon grease for 10 minutes, stirring occasionally. Add the onion, garlic, tomato paste, red wine, beef stock, salt and pepper. Cover the pot and simmer for 40 minutes. Add the mushrooms. Simmer for another 20 to 30 minutes. The venison is done when the meat falls off the bones. Remove the bones.

Blend in vinegar, sugar and basil to taste. Add sour cream before serving.

Serve with applesauce, boiled potatoes with freshly chopped parsley, or mixed rice (¾ white and ¼ wild). Serves 6 to 8.

Leg of Venison in Mustard and Pepper Sauce

approximately 5-lb. leg of venison
5 oz. lard, cut into 2 x ⅛ x ⅛-inch strips
1 tbsp. mustard powder
1 tsp. black peppercorns, crushed
salt to taste
4 oz. margarine
1 big onion, coarsely chopped
1 carrot, coarsely chopped
1 parsley root or parsnip, coarsely chopped
1 cup beef stock (from concentrate)

MARINADE
1 carrot, coarsely chopped
2 medium onions, coarsely chopped
1 tsp. black peppercorns, crushed
1 cup red wine vinegar
1 cup red wine
¼ cup brandy
1 cup beef stock (from concentrate)

SAUCE
3 oz. honey bread (available in supermarkets) or Nuremberg-style Lebkuchen, ground
6 tbsps. marinade
4 tbsps. red currant jelly
1 tbsp. mustard powder

Rinse the leg of venison in cold water. Remove the skin. Mix all ingredients for the marinade in a deep bowl. Submerge the venison completely (if necessary, add water). Cover with a lid and store the venison in a cool room for at least 5 hours.

Remove the venison from the marinade. Save 1 cup of the marinade. Lard the leg of venison across the grain fibers with the strips of lard. Rub with mixture of salt, crushed pepper and mustard powder.

Heat the margarine in a pot and brown the venison for 10 minutes, flipping it over occasionally. Add the onion, carrot and parsley root or parsnip. Brown for 5 minutes. Add the beef broth, cover with a lid and roast in preheated oven (on the lowest rack) for approximately 90 minutes at 350°F. Add water as moisture evaporates. Remove the venison from the pot and place on warm serving plate.

Strain all of the pan juices and mix it well with the ingredients for the sauce. Add salt, pepper or vinegar to taste. Bring to a short boil. Pour ⅓ of the sauce over the leg of venison and serve the remainder in gravy boat.

Serve with potato dumplings and Boston lettuce. Serves 6 to 8.

The Best Potato Dumplings Ever

2 lbs. potatoes
1 cup all-purpose flour
¼ cup farina
2 eggs, run in, and
1 egg yolk, beaten
½ stick margarine, soft
salt to taste
4 tbsps. fresh parsley, chopped

Boil the potatoes, then peel them (you can do this part of the recipe as much as a day in advance).

Fill the stockpot with 4 inches of water and bring to a boil. Mash the potatoes manually with a ricer. (Don't use a food processor.) Mix in the remaining ingredients, except the parsley. Let it sit for 10 to 15 minutes (don't wait longer because the mixture will liquefy).

Dip your hands in flour and form about 8 to 10 dumplings. Submerge dumplings in the hot water and simmer for about 10 minutes. They are done when they float to the surface. They're best when served right out of the pot (or keep them warm in a covered bowl). Top them with the parsley. Serves 6 to 8.

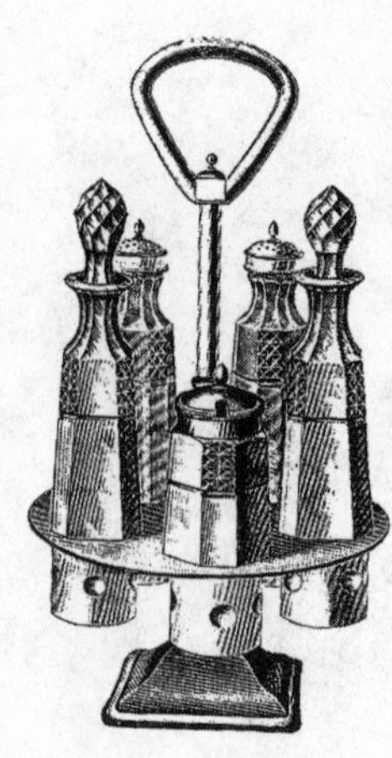

Corn Cakes

12 oz. corn kernels, cooked
1 ½ sticks unsalted butter, clarified
1 tbsp. red peppers, blanched and chopped
1 tbsp. green peppers, blanched and chopped
1 cup half-and-half

5 eggs
3 oz. flour
salt to taste

Sauté the corn in butter. Add the red and green peppers. Set aside to cool in a bowl, then mix with half-and-half, eggs and flour to make a batter. Add salt to taste. Form into small pancakes (1 ½ inches in diameter), and fry them in hot butter.

Good Venison, Great Wine

Should venison be aged? Should it be marinated? Are young bucks more tender and tastier than old ones? Should gaminess be toned down? And what is the best way to cook venison?

All these questions are very much interrelated and all have to do with the amount of fat on the animal. The simple rule is, the more fat, the tastier and more tender it will be. A skinny young doe won't taste as good as a fat old buck, but a well-fed young animal is preferable to a wizened old one. Venison does not have the kind of marbleizing streaks of fat that run throughout a steer's flesh. Instead the fat lies on top of the flesh, which adds succulence as well as flavor, although fat should be trimmed from steaks and chops because it doesn't look particularly appetizing. When you do find an animal without much fat you may have to lard it with salt pork, especially if it is to be roasted.

✻ ***Tender Care:*** Too many cooks believe the purpose of marinating is to kill off the strong gamy taste. The fact is, wild game is rarely gamy, no matter how old the animal is, unless the meat has been mishandled or aged incorrectly. Marinating will cover up some strong flavors, but the best venison, wild or farm-raised, should not have a strong flavor to begin with. If you wish to add flavor, fine, throw some peppercorns, shallots, thyme, savory, chervil or juniper berries into the marinade, but don't try to mask the flavor of the meat itself.

A good marinade is intended to break down the fibers of the meat, rendering it more tender. For that you need an acid, the best being a great good wine. You need not spend $50 on a fine Burgundy, but a

$10 to $15 bottle of hearty Chianti, Côtes-du-Rhône or a big California cabernet or zinfandel will make a very good marinade. But unless absolutely necessary, don't marinate venison at all.

Tenderness can be best achieved by aging, because just-killed deer are always going to be tough. The carcass must cool down and muscle tissue needs time to relax. Tradition dictates that it be hung by the head outdoors in cool weather (35°F to 40°F), for anywhere from three days to a week or so. But those worried about leaving game outside can just as easily tenderize meat in the refrigerator or freezer.

Finally, what cuts are suitable for which methods of cooking? The loin and loin chops are great for grilling over an open fire, as you would any beef steak. Roasting is better for rump and sirloin tip, and long braising is required for tougher cuts like shoulder.

However you prepare it, keep it simple. And if you have used a wine marinade, drink the same wine when you sit down to eat. Otherwise, get the best, most expensive red wine you can afford. Great venison demands a great wine, and nothing will ever taste better.

Mike Toth's Venison Lollipops

Cut venison tenderloin (backstrap) lengthwise, leaving ½ to 1 inch uncut. Open it up (you'll have a rectangle of flat meat) and pound with a tenderizer hammer or your fist. Be careful not to break or punch holes in the meat.

Roll the meat into a cylinder and place wooden skewers through it at every inch or so. Then cut between the skewers so you're left with large lollipops of rolled-up venison.

Sauté ½ cup of finely chopped onions in about 6 tablespoons of oil (use a fairly large saucepan). Then add the following:

1 cup lemon juice
⅔ cup vinegar (wine vinegar is best)
½ cup ketchup
½ cup water
2 tbsps. brown sugar
2 ½ tsps. salt
½ tsp. black pepper
2 tsps. dry mustard powder
1 ½ tsp. red mustard powder

Bring it to a boil; then let it cool and transfer it to a big bowl. Marinate the lollipops for at least 1 hour (overnight is best).

Grill the lollipops over a hot grill for only a few minutes on each side—the marinade will have "cooked" the venison some already.

(You may want to double or even triple the marinade recipe if you're using a large tenderloin, or if you want to add marinade to the lollipops while they're grilling.)

Clean Cuts

One of the problems with butchering venison is doing a neat job of cutting the steaks and chops. The meat quivers like a jellyfish, and making a good straight cut is difficult. One of the best ways to solve this problem is to add an end board to the cutting board. With slight hand pressure on the meat near the end board, you'll be able to get neat, straight cuts.

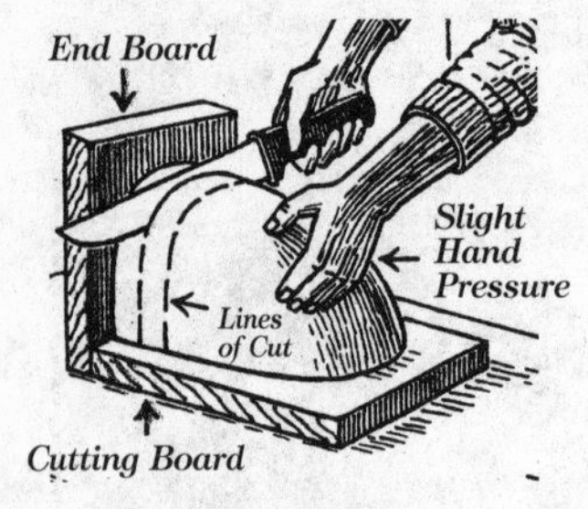

Just the Backstraps, Please

These two strips of meat run along the top of the backbone, just under the skin, and many people call them the tenderloin or top loin. By whatever name, backstraps are choice pieces of meat, and they are often used whole for cooking Wellington-type recipes. They are sometimes left attached to the backbone and taken out as a unit, called a saddle, which can be used in your favorite roast recipe. Usually, however, the backstraps are cut out and sliced into small steaks, chops or cutlets. These can be beaten with a meat mallet, floured and fried, or they can be marinated and grilled. Better yet, make them into pepper steaks, as follows:

backstrap steaks, ¾ inch thick
olive oil
garlic clove
peppercorns
salt to taste

Beat the steaks with a meat mallet or the edge of a plate. Marinate them lightly in olive oil containing a minced clove of garlic. After 1 hour or longer of marinating, drain the steaks and pat them dry. Grind some peppercorns and press the coarse pepper bits into each side of the steaks. Let them sit for 10 minutes at room temperature. Preheat the broiler. Place the steaks on the uppermost broiler, so that the surface of the meat is within two inches of the heat source. Broil the steaks for 4 minutes on each side. Do not overcook—the meat should be pink inside. Salt to taste.

HE IS DEAD

to his own interest, who deals in Arms and Ammunition, or who shoots a Rifle, Pistol, or Shot Gun and has not a copy of the IDEAL HAND BOOK, No. 3, 64 pages of solid information. Just out.

BE ALIVE and send for one, **FREE.** Stamps for postage will not provoke us.

IDEAL MFG. CO., Drawer 86 D. New Haven, Conn, U. S. A.

THE FRESHEST, CLEANEST AND BEST.

A MAGAZINE that has no rival in the field that is particularly its own.

With ample means to accomplish its plans of future improvement, and with the assistance of the brightest corps of contributors that can be gathered between the Atlantic and the Pacific, SPORTS AFIELD will strive to deserve, hereafter, all the patronage and kindly commendations that it has won in the past. Send Two Dollars for a year's subscription.

SPORTS AFIELD PUBLISHING CO, DENVER, COLO.

STEVENS SPECIAL FIRE ARMS.

Wherever shooting is done, in all parts of the world, the **Stevens Rifles and Pistols** are being introduced, and on account of their superior accuracy are preferred.

STEVENS PISTOLS have made more brilliant records and have done finer work than any other pistol ever made.

STEVENS RIFLES have made records which were considered impossible.

STEVENS POCKET RIFLES are carried by ladies, anglers, tourists and hunters. They are marvels of accuracy, compactness and beauty.

STEVENS LADIES' RIFLE is the proper rifle for ladies. It is wonderfully accurate and has no recoil.

The 22 long rifle and the .25 rim-fire cartridges, the most accurate small-bore cartridges made, were originated by this company. These celebrated rifles and pistols are manufactured by

THE J. STEVENS ARMS AND TOOL CO.,

Send for Catalogue.

P. O. Box 5680. Chicopee Falls, Mass.

9

Miscellany

Last fall a party of Indians from the Flambeau Reservation came over on a hunting expedition, and while there killed a splendid white deer. It was a pure albino, with pink eyes, and weighed about 180 pounds. The Indians brought it into camp and, I believe, afterwards sold it for $20.

—R. P. Sharples, *Sports Afield*, July 1899.

The Days of All-Purpose Deer

Native Americans used every part of the deer they killed. The hide was used for clothing, shelter and rawhide. The hair removed during the tanning process was saved to insulate moccasins and papoose cradles. Bones were fashioned into needles, awls, hide scrapers, fish hooks, harpoon points and children's toys—after the bone marrow had been extracted and eaten.

Tongue, heart, liver and kidneys were all consumed and relished, and less edible organs such as intestines were used too. Lengths of intestine were washed, dried, and stuffed with pemmican (a mixture of dried, pulverized meat, fruit and rendered fat) for later use. The deer's stomach served as a water container or for storage of other foods. The bladder was likewise dried and used as an herb or medicine pouch.

Brains were commonly used to tan the animal's hide. Leg and back tendons were carefully removed and dried and saved as sinew for sewing, lashing arrowheads to shafts, backing hunting bows, making bowstrings and many other cordage purposes.

Fat from the animal's flank was rendered and used as a water-repelling agent on hunting tackle and clothing. It was also mixed with herbs and used as a medicinal salve and insect repellent.

The deer's teeth made fine jewelry, provided they weren't too old and worn. To make glue, hooves were boiled along with hide scrapings or any other scraps of soft tissue that were not otherwise of any use.

What about the antlers? They were usually reduced to stoneflaking tools, spear points, spoons and other household tools. It's doubtful whether many racks ever hung as mere trophies on the wall of an Indian lodge.

Scent Effectiveness Drops With Degrees

If you've wondered why your favorite buck lure doesn't seem as effective during late fall or winter, blame the temperature, not the lure. Tests have proved that liquid lures and scents are only 25 percent effective at 20°F. The evaporative effect of a liquid-based scent or lure is greatest at approximately 80°F and decreases to just 10 percent as temperature drops to around 10°F.

Unfortunately, human odor, wafting from a warm body, is not proportionately reduced as air temperature falls.

Warming deer scents and lures can restore their effectiveness, but use caution. Do not use an open flame or a source of heat that gives off an odor. The odorless flame-free type of hand warmer works nicely and lasts for several hours.

Tie a Tree Limb

When setting up a tree stand, don't saw or chop the branches away. Instead, tie them up with wire or cord. There are several advantages:

* Tying provides a denser background for you—deer are less likely to spot you.
* Tying is quiet, and you can begin effective hunting immediately. When you chop or saw, you normally have to wait an hour or more before deer begin moving through.
* A few feet of wire is lighter to carry than a hatchet or most saws. The wire or cord can also be used to lift your gun or bow into the tree and to drag your deer.
* Tying doesn't damage the tree—something landowners are certain to appreciate.

If branches are in your way, tie them back.

Make Hunting Friends Early

If local law allows you to hunt posted land with the owner's consent, it's never too early to make friends. Landowners like to accommodate hunters they know personally, but they often cut off hunting privileges when requests come in autumn. The ticket here is to plan ahead. A great way to build an honest, mutually beneficial friendship is to help with the spring and summer chores. Most farmers take a quick shine to someone willing to dig post holes, mend fences, or help with the haying. A weekend of hard work in exchange for a new friend and a season of deer hunting—now that's a hard-to-beat bargain for most any hunter.

Antler Repair

Sportsmen occasionally take whitetails with broken antlers, a poor token of what would otherwise be a gorgeous rack. It's not easy to find appropriate materials for a repair job, but it is possible. One method that works is to use gel-type rifle glass bedding, mixed with wood ashes. The ashes give the epoxy-resin mixture a perfect gray color and texture, removing the glare and matching the antler's shade and feel. This material also serves well as a bond where parts of an antler are cracked or broken.

Just follow the instructions found with the glass bedding material and stir in sufficient wood ash to get the right color. As the material nears the setting stage, sprinkle some extra dry ashes on the patch and rub the surface lightly with your finger until the color matches the antler. No special talent is required.

Get an Aerial View

Aerial photographs available from your local Agricultural Stabilization and Conservation Service office are valuable aids for deer hunting. They are most useful when combined with topographic maps to give you a more complete picture of the deer habitat you plan to hunt. The photo will show things such as clear-cut areas, pipelines and new trails or roads built since the topo map was drawn.

Determining Deer Weight

There is a method for estimating the weight of deer killed far from camp that is simple, easy and good for winning a bet or two from nonbelieving hunters.

First, purchase a 60-inch tape measure (it sells for about a dollar). Keep it in your hunting shirt pocket after marking the following formula on one side of the tape with a felt-tip pen: (C x 5.6) – 94 = live weight. "C" is the chest measurement. A small pocket calculator will fit in your shirt pocket with the tape and will let you figure the weight quickly and easily.

For example, a deer with a 42-inch chest girth, taken where the heart lies, will weigh about 141 pounds. Now that the live weight is known, multiply that figure by .78 to get the field-dressed weight, in this case about 110 pounds. That leaves only the freezer weight, which is approximately one-half of the field-dressed weight, or in this case, about 55 pounds.

Deer Hunting Is Tough

No one has to convince a veteran deer hunter of the ability experienced whitetail bucks have not only to stay alive, but to seemingly vanish into nothingness. However, it took a three-year Michigan study, under carefully regulated conditions, to prove just how difficult deer hunting actually is. The experiments started with 647 acres of conifer swamps—mixed hardwoods and pine barrens enclosed by a deerproof 11-foot-high fence. During the three years of the experiment, herd populations hovered between 26 and 34 members, making the area one of the most densely populated for its size in the state. About 20 percent were antlered bucks. The animals were kept under almost constant surveillance by game biologists.

Two years the hunts were in deep, wet snow that hampered the hunters. In only one year were the conditions good. Both buck-only and any-deer hunts were staged. The results were as predicted—deer

Deep, wet snow can dampen your chances of finding a deer.

hunting was tough. Hunters required 14 hours to shoot a deer during the deer season and 50 hours to down a buck in the antlered hunt. Trying to shoot a buck during any-deer days was highly unprofitable—140.5 hours were required to do it. Most astonishing was the fact that hunters were able to see—much less shoot—only a tiny percentage of the deer known to be in the enclosure. A mere 2.1 percent of the herd was sighted on an average day. In one experiment six hunters took 15 ½ days to even sight a buck in the one-square-mile area where seven antlered bucks were known to dwell.

In another similar experiment in South Dakota, a buck outfitted with a beeper radio was released, and five supposedly expert hunters were sent out to find it. After seven days of failure, the hunters were directed to the exact area where the radio direction-finder indicated the animal was. Again, failure. Not believing their eyes and ears, the hunters were sent back to the exact location of the radio signal. The buck was finally routed out—but only after one hunter accidentally stepped on it lying flat on its belly in dense underbrush!

How to Read Deer Hair

Deer hair, as it is dislodged or cut off by bullet or broadhead, is perhaps the least-used and -understood information available to the hunter. Good deer trackers have learned that deer hair, which often tells wound locations, is sometimes the most valuable clue they have.

The type and color of hair can tell an alert hunter where it came from. The amount or condition of the hair can also indicate the type of wound as well as its position. Both bullets and broadheads making a solid, direct hit leave far less hair than severely angled or grazing shots do. It's easy to make a chart of hair samples from various parts of the deer. It's important to use hair in fall pelage. Fold a light-colored, firm piece of cardboard in half. One half holds numbered specimens and the other is a legend reference. Use double-coated tape or a liquid adhesive to retain the hair samples. Pull the samples out, rather than cutting them, to show full strands.

Then study and try to memorize the chart, noting the differences in color as well as multicolor distribution, length and overall makeup from different body areas.

Mellow, Yellow Deer

California has a population of more than 500 fallow deer—a yellowish-brown deer with many white spots on the body. One hundred of these deer, originally from the Mediterranean, were introduced by the Hearst Ranch in about 1940.

The Coues' Connection

If you're not familiar with the Coues' deer (technically pronounced *cows*—referring to Elliott Coues, the 19th-century naturalist for whom the deer is named—although absolutely everyone says *cooz*), it is the wee desert version of the white-tailed deer found in southern Arizona and New Mexico, as well as in northern parts of Sonora and Chihuahua states in Mexico. How "wee"? Any buck with an inside spread of 12 inches is worth bragging on, and a deer that dresses out at more than 100 pounds is a very big Coues'. And yet—though admirers of the Coues' claim "it's a whitetail that lives in mule deer country and thinks it's a bighorn sheep"—it remains very much a white-tailed deer, spending, like most whitetails, all its life in the same square mile or so in which it was born.

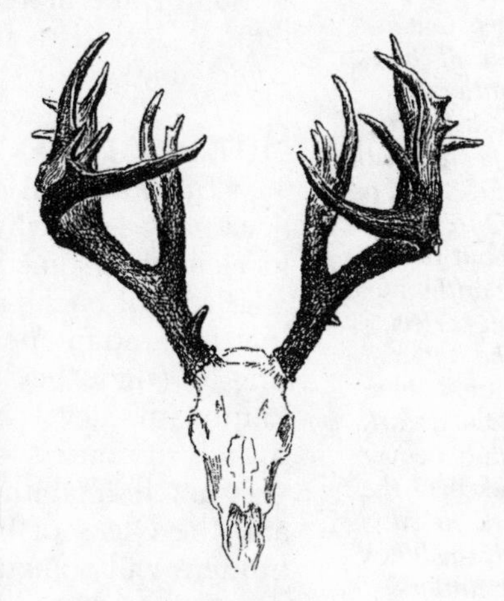

Big Antlers, Big Deer: Serious trophy hunting (as opposed to blind luck) means serious scouting well before the season opens. To determine if an area holds trophy deer, the first thing you should do is look for big antlers. As soon as the deer have shed begin your search for sheds. Look around places where there's food and water, expanding outward from there, and work your way along deer trails. When you finally find a trophy-sized shed, you can be fairly confident the buck that dropped it is nearby.

Then start to look for the deer itself. A trophy buck usually does not reach its full potential before age 6, and from ages 6 to 10 the deer may maintain its antler size. You may spend several seasons following a particular deer, collecting its sheds each spring, identifying the antlers by the size and shape and grain texture of the base, using them to chart its growth. You might also look at the frame and pattern of the antlers, and as soon as you spot a buck in velvet that fits those characteristics, halt your hunting through that deer's territory and back off to a vantage point from which to glass without further disturbing the deer.

Study the deer to learn its daily routine of feeding and bedding, observing it right up until the opening of hunting season. Know that many true trophy bucks never really develop routines—this could account for why they live to be trophy bucks—but may simply remain bedded all season in thick cover where they have food and water. Such a situation may mean that you have to sit and wait for movement.

Sound like more work than it's worth? Well, perhaps it does. But nobody ever said that trophy deer come easy.

Bleached Antlers

The next time you bag a buck and don't know what to do with the antlers, try this: Use a hacksaw to cut off the top of the skull, antlers attached. Skin it, then clean off any flesh. Set the antlers in a place with a lot of exposure to the sun. It will take about two years, but the antlers will bleach out to a perfect white. After the antlers have cured, mount them on a board stained dark mahogany. Glue a covering of black felt over the skull. The dark board combined with the black felt accent the antlers' bleached-out whiteness, making a beautiful wall mount.

Summer Deer Hunts

✷ ***Blacktails:*** Along the Pacific Coast, black-tailed deer may be bowhunted as early as July in certain locales, with some rifle hunts beginning in August.

Blacktail bucks at the start of the season are still in velvet and frequently travel in bachelor herds, splitting up when they strip their antlers. Since the conjoining of "level ground" and "blacktail hunting" would, of course, qualify as an oxymoron (the deer very much inhabiting the ups and downs of oak draws and the edges of heavy timber), one technique that hunters will sometimes employ is an easygoing one-man drive—more like an "urge"—to push deer out of rugged canyons toward a stander.

Summer deer hunting, everywhere, is also extremely hot work, with daytime temperatures in blacktail country ranging from 85°F to 100°F. Deer in velvet that are still building antlers will spend a lot of time feeding, but on those real scorcher days, deer tend to move in the early mornings and late afternoons. Expect to spot them bedded in shade at midday, then be prepared to stalk, watching for rattlesnakes and poison oak. For summer hunters, brown Bushland camo will help conceal your approach.

Antler Expertise

Visit a taxidermist's shop and study the various racks on display to learn what qualities make an exceptional set of antlers. Also study the Boone and Crockett scoring system so you'll know what factors make a rack a trophy.

✷ ***Mule Deer:*** From sagebrush to aspen-and-spruce high country, summer mule deer often can be found in slightly cooler climes, but only slightly. New Mexico, Utah, Colorado and other states throughout the Rockies have bow seasons beginning in late summer. For warm-weather mule deer bowhunters, setting stands along travel routes to and from hayfields is a good strategy. And in dry years, waterholes are also good spots to stake out. But mule deer at this time—still in velvet and sometimes in bachelor herds of three to five—are also less wary than they may be later in the season, and thus are easier to stalk. In early September the brush is still leafed out, and a

Summer mule deer hunters tend to have success along travel routes.

green-background camo would be recommended, but remember that quiet matters more than pattern. And like all summertime deer, a mule deer should be skinned and hung immediately after killing, capes fleshed and salted ASAP.

✳ ***Whitetails:*** In late summer, deer's natural feed has desiccated and turned tough under the hot sun; so the animals head for the tender shoots and leaves in green, cultivated fields.

Summer whitetails often feed in cultivated fields.

Both archery and gun hunting are open in certain spots—particularly in the South—and from a stand a bow hunter has a chance to take a trophy velvet-antlered deer as it moves to and from food plots. The rifle hunter's also likely to spot a trophy buck, and the very best time for doing that is in the last 20 minutes of shooting. Lengthy shots can be the order of the day then, so it's wise to know your bullet and load intimately. Good optics, in your choice of both scope and binoculars, are also important for summer deer hunting, because so much of it involves being able to spot the deer well ahead of the shot. Remember, though, that you want your binoculars and scope to be of equal quality, because first-rate binoculars that let you spot a distant deer in fading light do not serve you well when you are then unable to find that same deer in a third-rate scope.

Along with plenty of water for the 100°F daily heat, a summertime whitetail—or blacktail or mule deer—hunter should also carry insect repellent. An annoyance? Sure. But to gain a two-month jump on deer season, and beat those summertime blues, it's certainly a small price to pay.

No More Lost Deer Tags

One of a hunter's most embarrassing moments is trying to explain to a skeptical game warden that his deer tag "must have fallen off." To avoid this problem, strengthen the tag by adding several strips of reinforced transparent tape to both sides. Be sure to leave room for your signature and address. Punch a hole in the taped section, and fold a 12-inch piece of fine copper wire into the license. After your deer is down, wrap the tag securely around an antler with the copper wire to hold it in place.

The Boone and Crockett Legacy

In 1887, two very important beginnings took place. One was the start of an outdoors publication called *Sports Afield*. The other was the establishment of a club that would become one of the most influential forces in natural resources management.

Although few in number, those men who gathered for a dinner in New York City, at the invitation of Theodore Roosevelt, were giants in a wide spectrum of American life. They were artists, politicians, military leaders, publishing magnates, explorers, writers and scientists. They were visionaries who could and did look beyond the resource exploitation binge of a young, aggressive nation to see the needs that lay ahead. Equally important, they had the drive, the skills, the tenacity and the clout to effect massive changes that will benefit America forever. That night they founded the Boone and Crockett Club.

What brought and held these men together? What was the tie that bound? All were avid big-game hunters. The initial function of the club they formed was to take care of the big-game resources of North America.

For most hunters today, Boone and Crockett bring

Buckskins for Britain

Deer hides were once a significant item of commerce between the American colonies and England. Old customs records show that 2,601,152 pounds of deerskins were shipped to England between 1755 and 1773, just from Savannah, Georgia.

to mind a big-game scoring system, a method of comparing North American trophies. "It made the Book" is a line that is distinctive and stands alone. It speaks of the Boone and Crockett record book and refers to an animal that scores high enough for inclusion in that publication. "It'll make B&C" and "It's a Boonie" are integral phrases of hunting's language.

The B&C record book is the standard that most hunters use in judging North American big-game animals, but there is vastly more to this venerable organization than being merely a keeper of "the Book." It remains one of the most effective conservation groups in the land, and its membership of 150 bears witness to the fact that a few people can make a difference.

Conservation! We take the word for granted, but when Teddy Roosevelt instigated the formation of the B&C Club, the concept of "wise use" of natural resources had not really been developed in this country. One of the first advocates of this position was Gifford Pinchot, a B&C Club member who, while serving as head of the U.S. Forest Service during Teddy Roosevelt's administration, worked with the president to define the complex actions required to manage and maintain natural resources.

Boone and Crockett. The words have such a lilt that many of us seldom think of whom they refer to: Boone—as in Daniel Boone; Crockett—as in Davy Crockett. The men who gathered in 1887 selected those names—those men—as emblematic of the spirit they wanted to convey. Daniel Boone and Davy Crockett represented a reaching out to the wild places of the West; they were pioneers and explorers. They were, above all, big-game hunters.

The prime mover in B&C in the early years was George Bird Grinnell, whom many consider the father of conservation in America. Pushing the club and using his podium as editor of a weekly newspaper called *Forest and Stream*, he successfully advocated a wide range of reforms in both law and hunting philosophy. The spirit of the B&C Club became—and remains to this day—a code of ethical conduct for the hunting experience.

The current B&C "Rules of Fair Chase" reflect that code, prohibiting: 1) spotting or herding game from the air and then landing in its vicinity for pursuit, 2) herding or pursuing game with motor-pow-

Make a Map File

If you do much traveling, you know how maps soon become a jumble. You'll never get rid of this problem unless you start a map file. Simply fold each map to the same size, and mark its upper right corner with the area it covers, its source, and the type of map it is. For example, one map might read Astoria, OR; Geological Survey; topo. After they are marked, file them alphabetically by state first, county second and specific section third. Keep them in marked slots in a bookcase, or, if your collection is small, in an inexpensive expanding file.

ered vehicles, 3) using electronic devices to attract, locate or observe game, or to guide the hunter to such game, and 4) hunting game that is confined by artificial barriers (including escape-proof fencing) or transplanted solely for the purpose of commercial shooting.

The influence of B&C was dominant in a wide range of projects, including passage of legislation establishing critical national parks and instituting methods of governing and protecting them. Club members founded the New York Zoological Society and the Bronx Zoo, which cooperated with B&C in the stocking of bison on national preserves; this action may well have saved the species from extinction. It originated the concept of national wildlife refuges, was the godfather of the American Wildfowlers (predecessor of Ducks Unlimited), pioneered the creation of the National Forest System and brought about both an end to legal market hunting and passage of the Lacey Act, one of the most important pieces of conservation legislation in history. That act made it a federal offense to transport across state lines an animal taken in violation of state laws, and it remains one of the most potent weapons in the fight against abuse of wildlife resources.

Today, B&C continues its record-keeping program, awards grants to support research in wildlife ecology, and periodically sponsors symposiums on various big-game animals. In early celebration of its 100th year, the club bought a superb 6000-acre tract in northwestern Montana that was renamed the Theodore Roosevelt Memorial Ranch, and established the Boone and Crockett Club Foundation to manage it. Harboring a wide variety of big and small game, and wintering what may be the largest herd of elk and deer along the east slope of the Rocky Mountains, the ranch is dedicated to one of the most serious problems facing conservation: how to manage the inherent conflict between wildlife preservation and man's use and development of the land.

Cut Firewood for Deer

When cutting your winter's supply of firewood, two things will benefit your local deer population. Always leave the tree crown lying on the ground rather than hauling it away. This will provide deer with browse during the colder months. Never cut the standing tree trunk flush at ground level. Always leave a three-foot stump. Aside from being easier on your back (you don't have to bend over when using your saw), the stump will sprout "sucker" shoots the following spring, providing deer with tender buds, stems and leaves to nibble on.

A Bit of Buckskin

Every year thousands of deer hides are tanned by smart hunters who know the superior value of this material for making many things, from shirts to moccasins. One excellent use for a piece of leftover buckskin is to make a nice thin wallet, and it takes no special skill to do.

Lay a rectangular 10 x 4-inch piece of leather on a flat surface. Bring up the end about four inches. This leaves an overhang of two inches, which will be the flap of the wallet. Now the wallet can be sewn. Soft leather can be sewn on a machine. Heavier leather may require an awl, or the aid of a shoe-repair shop (usually at minimal cost). The stitching goes up the two sides only, which immediately produces a wallet. The unsewn portion on top flops over for a cover.

This thin wallet of buckskin will hold all of the credit cards that a busy man of the world might need, plus a few dollar bills and any other essentials.

Dressed-Up Antlers

Deer antlers accumulate in strange places—in garages, tool sheds, attics, under the house and eventually some end up on a forlorn fence post. Maybe they are just forgotten altogether and are left to collect dust or become sun-bleached in the backyard. Each year more are added to the pile, those trophy memories, not quite worthy of a head mount.

These lesser trophies have value, however, not just for mounting on a plaque board, which never seems to get done. They are unique for making buttons, necklaces, belts, knife handles and good-luck charms. Cut up my hard-earned trophies, you say? Well, of course, some racks you just can't part with, but what of that lowly forked horn you bagged that you're just not proud of?

Antler is bone, and this makes it a material that is hard, useful and long lasting. It can be shaped, drilled sawed and sanded, much the same as wood, into bear-claw type necklaces, good-luck charms, key-chain ornaments, knife handles and even gear-shift knobs. Place the antler in a vice, measure two inches from the tip or tine of the point and cut off with a metal-cutting hacksaw. Drill a hole in the longer end and string with a bead chain or a leather thong.

For buttons, start near the end where the tip was sawed off or go down to near the base of the antler for larger pieces. Measure the desired thickness and saw off. These pieces can be drilled two times or more in the center for attachment to garments. They can be sanded smooth or left rough. You may want to change the buttons on all your coats or make a headband or maybe a belt, strung with heavy leather string.

Deer on the Road

The next time you whiz past a deer-crossing sign on the highway, you might want to slow down, perk up and pay attention. Each year more than 300,000 deer are hit by motorists, and while that number is difficult to confirm, it could be much higher. According to a Cornell University study on auto collisions with deer in New York State in 1989, six deer were hit by a motorist for every one reported. Take 1993 as an example: The New York Department of Environmental Conservation recorded 10,332 reported collisions. That means the actual number of deer hit was probably about 62,000.

In other states, the numbers are more startling. In 1993, Pennsylvania recorded nearly 46,000 collisions, and Michigan, with another of the highest rates in the country, recorded 47,813. You would almost think that motorists were trying to outdo traditional hunters, except for the fact that it cost them about $1500 for every deer they hit. The collisions have also resulted in close to 100 human lives lost every year.

The peak season for these collisions coincides with the rut, the most activity occurring in early December. Keep in mind that there is less chance of injury if you hit a deer than if you swerve to miss it and hit something else. Also remember that deer will exhibit complete confusion when confronted with cars, behaving more like squirrels than dogs or cats; they will try to outrun cars rather than dodge them; and they are most active at dawn and dusk, when commuter traffic is heaviest.

When a Buck Was Worth a Dollar

Certain words in our vocabulary stem from the nation's frontier days. For instance, the word "buck," which we commonly use to describe a dollar bill, was coined by early woodsmen. Prior to the American Revolution, buckskin was a valuable trade item. According to historians, buck deer hides, prized for the leather that made durable britches, were as good as hard cash. For early settlers, one buckskin would buy a dollar's worth of salt, gunpowder and other vital necessities, and that's how the words "buck" and "dollar" came to mean the same thing.

Putting Deer on the Pill

Wildlife biologists have tried for over 20 years to produce an alternative to hunting as a means of controlling deer populations, most notably contraception. But after extensive study and research, they are still looking for a solution that is cheap, convenient, effective and safe. So far, nothing has turned up that makes sense on a large management scale, and so far, nothing beats hunting.

In national parks, wildlife refuges and suburban areas, however, hunting is not always an option, and the rapid spread of deer in these areas has reached crisis proportions.

To solve such problems, scientists have been testing a process called immunocontraception, in which a small dose of a drug called PZP (made from a substance found in pig ovaries) is used to inhibit fertility. But the problem of getting such a drug into hundreds of thousands of free-ranging deer remains. Dart guns have been suggested as a means of delivery, but as any hunter can tell you, it's not easy to get one deer in your sights, let alone an entire herd.

The Elusive Whitetail

This short, historical item is dedicated with sympathy to all those who pursue the popular white-tailed deer. One of America's first national parks was a 92-square-mile area in southwestern Oklahoma. Now named the Wichita Wildlife Refuge, it was the hunting grounds of the Apache, Comanche, Kiowa and Wichita Indians. With the white man's

The white-tailed deer is an incredibly resilient species.

arrival, market hunters turned on the animals with terrible ferocity. One by one, game species became extinct. Bison were wiped out. Then the elk went. Wild turkeys disappeared. The last antelope fell. Even longhorn cattle were shot to extinction.

When conservationists started speaking out, the animals began to receive protection and a warden force was established, and only one shy and delicate creature was found to have survived the decimation. Out from the nooks and crannies of the range, from places where one would think not even a mouse could hide, came—you guessed it. Only the white-tailed deer had been smart enough to survive.

Trophies in the Attic

It may not even be hunting season yet, but you could still "bag" that record-book buck . . . from right over the fireplace. Roger Sellner, who scores racks for the Boone and Crockett Club, says now is a great time to find that trophy head. Don't laugh. You might just have a record-book buck in the attic and not even know it.

Several years ago at the Oregon Sport Show, a young man brought in a 40-some-point nontypical head of a mule deer his grandfather had shot. It had hung over the ranch fireplace for 65 years and was completely blackened, with the hide scorched. But he had the original photos of his grandfather with the deer. The head wound up the new Oregon record at 321.1 points, and is No. 4 in the world.

White-tailed deer

Appendix

The following deer-hunting tables have been reprinted from the 1996 Deer Hunters' Almanac, *which is published annually by* Deer & Deer Hunting Magazine, *700 E. State Street, Iola, Wisconsin 54990-0001; 715/445-2214. Season dates and regulations change frequently, so contact your state fish and game agency for current information.*

Firearm-Hunting Statistics

State	Season Dates	Season Days	Resident Hunters	Nonresident Hunters	Deer Harvested	Rifles Legal?	Handguns Legal?
Alabama	11/20–1/31	83	205,000	20,500	270,000	Yes	Yes
Arizona	10/29–12/31	46	63,066	3670	17,519	Yes	Yes
Arkansas	11/13–12/18	36	Unknown	Unknown	79,385	Yes	Yes
California	8/13–10/9	58	303,720	Combined	27,553	Yes	Yes
Colorado	8/28–12/31	122	96,500	61,100	54,100	Yes	Yes
Connecticut	11/21–12/10	18	24,000	1100	6629	Yes	No
Delaware	11/11–1/21	11	17,450	1236	5872	No	No
Florida	10/30–2/16	72	184,910	3061	Combined	Yes	Yes
Georgia	10/22–1/8	79	330,000	20,000	345,100	Yes	Yes
Idaho	9/15–11/20	65	142,000	16,877	61,200	Yes	Yes
Illinois	11/18–1/15	13	143,308	Combined	92,000	No	Yes
Indiana	11/13–11/28	16	165,000	4000	74,800	No	Yes
Iowa	12/3–12/18	16	131,188	700	61,663	No	No
Kansas	12/1–12/12	12	52,604	Just Started	30,000	Yes	Yes
Kentucky	11/12–11/21	10	141,960	2839	68,000	Yes	Yes
Louisiana	10/30–1/17	68	189,100	2142	189,100	Yes	Yes
Maine	10/30–11/27	25	160,250	31,300	26,608	Yes	Yes
Maryland	11/26–12/10	13	106,000	14,000	33,785	Yes	Yes
Massachusetts	11/28–12/10	13	70,000	1350	6300	No	No
Michigan	11/15–11/30	16	738,200	17,800	251,410	Yes	Yes
Minnesota	11/6–11/26	21	443,000	8500	188,109	Yes	Yes
Mississippi	11/20–1/19	47	175,000	16,858	220,000	Yes	Yes
Missouri	11/12–11/20	9	436,341	10,390	154,159	Yes	Yes
Montana	9/15–12/31	36	143,100	15,000	153,928	Yes	Yes
Nebraska	11/12–11/20	9	53,967	1621	30,767	Yes	Yes
Nevada	9/11–1/2	97	45,000	10,000	5991	Yes	Yes
New Hampshire	11/10–12/5	26	66,872	10,028	6643	Yes	Yes
New Jersey	12/6–1/22	9	108,116	2358	27,197	No	No
New Mexico	10/17–11/15	22	59,000	5100	16,820	Yes	Yes
New York	10/22–12/13	53	600,000	30,000	Combined	Yes	Yes
North Carolina	10/17–1/2	67	260,000	14,000	150,000	Yes	Yes
North Dakota	11/4–11/20	17	89,691	669	65,375	Yes	Yes
Ohio	11/28–12/3	6	350,000	3000	104,400	No	Yes
Oklahoma	11/19–11/27	9	151,002	606	40,033	Yes	Yes
Oregon	9/3–12/18	91	233,100	2086	64,000	Yes	Yes
Pennsylvania	11/29–1/22	42	1,044,634	75,777	351,544	Yes	Yes
Rhode Island	11/26–12/4	9	7758	200	313	No	No
South Carolina	8/5–1/2	141	147,500	29,500	Combined	Yes	Yes
South Dakota	9/18–12/19	56	Combined	Combined	63,100	Yes	Yes
Tennessee	11/20–1/9	34	190,200	5228	85,568	Yes	Yes
Texas	11/6–1/30	85	588,000	12,000	456,000	Yes	Yes
Utah	10/17–10/23	7	97,000	9500	21,822	Yes	Yes
Vermont	11/12–11/27	16	89,423	17,082	10,043	Yes	Yes
Virginia	11/21–1/7	42	278,562	14,307	158,361	Yes	Yes
Washington	9/15–11/20	48	168,156	1083	47,701	Yes	Yes
West Virginia	11/21–12/17	21	329,472	46,065	132,820	Yes	Yes
Wisconsin	11/20–11/28	9	639,964	26,606	217,584	Yes	Yes
Wyoming	9/10–11/30	52	72,272	50,117	69,415	Yes	Yes
TOTAL			**10,131,386**	**619,356**	**4,542,717**		

Deer-Harvest Statistics

State	Estimated Deer Population	State	Total Deer Harvest	State	Resident Hunters	Nonresident Hunters	Season Bag Limit
Texas	3,525,000	Texas	476,000	Pennsylvania	1,407,822	29,000	1/day
Mississippi	1,750,000	Pennsylvania	408,557	Michigan	1,248,770	4351	2
Alabama	1,500,000	Michigan	374,640	Wisconsin	857,756	Unknown	3
Michigan	1,500,000	Georgia	345,100	New York	803,000	2055	2
Pennsylvania	1,178,368	Mississippi	301,000	Texas	666,000	71,500	1
Minnesota	1,000,000	Alabama	295,000	Ohio	649,000	3350	11
Wisconsin	1,000,000	Wisconsin	270,592	Missouri	541,488	2241	2
Georgia	950,000	New York	220,288	Minnesota	516,000	3813	2/day
Montana	910,000	Louisiana	214,900	West Virginia	508,868	23,212	5
New York	900,000	Minnesota	202,000	North Carolina	456,000	21,109	1
North Carolina	900,000	Virginia	201,122	Georgia	445,000	Combined	1
Virginia	900,000	North Carolina	177,000	Virginia	400,452	5725	10
Louisiana	875,000	Missouri	172,141	Tennessee	366,902	1200	3
Florida	819,420	West Virginia	169,014	California	337,661	Just Started	4
West Virginia	800,000	Montana	153,928	Indiana	302,900	5639	2
Tennessee	780,000	South Carolina	142,302	Mississippi	297,000	2807	6
California	760,700	Tennessee	138,615	Alabama	290,000	32,635	1
Missouri	760,281	Ohio	135,000	Kentucky	271,600	26,283	6
South Carolina	750,000	Illinois	113,000	Oregon	259,655	1928	2
Arkansas	700,000	Arkansas	110,401	Oklahoma	246,664	28,153	4
Oregon	686,600	Indiana	101,250	Louisiana	244,097	9658	1
Colorado	600,000	Kentucky	95,300	Florida	229,612	28,610	8
Wyoming	460,000	Florida	81,942	Illinois	227,325	11,934	4
Kentucky	400,000	Iowa	78,000	Washington	199,330	23,000	4
Washington	398,000	Wyoming	70,450	Maryland	199,000	2369	2
Ohio	350,000	North Dakota	70,293	Iowa	184,353	10,520	1
South Dakota	340,000	Oregon	70,000	New Jersey	178,801	16,664	2
Oklahoma	322,500	South Dakota	66,800	Idaho	177,600	4796	20+
Kansas	320,000	Colorado	61,500	Maine	177,250	6433	1
Iowa	300,000	Idaho	61,200	South Carolina	172,500	35,900	6
Indiana	290,000	Oklahoma	58,125	Montana	160,000	24,554	5
Maine	275,000	Washington	55,297	Utah	135,407	1338	6
New Mexico	260,000	Maryland	51,209	Vermont	125,821	6000	3
Utah	250,000	New Jersey	49,942	Colorado	117,300	1211	5
Arizona	243,000	California	45,000	New Hampshire	105,284	3394	1
North Dakota	230,000	Nebraska	37,277	North Dakota	101,737	104,182	8
Nebraska	210,000	Kansas	36,600	Massachusetts	100,000	650	8
Maryland	200,000	Utah	30,733	Arizona	82,708	34,500	10
New Jersey	150,000	Maine	27,402	Wyoming	80,383	4100	32
Nevada	149,000	Arizona	18,565	South Dakota	75,000	15,684	20+
Vermont	105,000	New Mexico	17,597	Nebraska	72,963	14,000	6
Massachusetts	70,000	Vermont	13,333	Kansas	70,179	11,910	1
New Hampshire	67,662	Connecticut	10,360	New Mexico	69,041	24,263	3
Connecticut	65,000	New Hampshire	9889	Nevada	48,154	20,233	3
Delaware	20,000	Massachusetts	8200	Connecticut	43,000	1307	1
Rhode Island	5500	Delaware	7424	Delaware	31,527	71,490	7
Idaho	Unknown	Nevada	6276	Rhode Island	16,058	32,141	1
Illinois	Unknown	Rhode Island	1322	Arkansas	Unknown	52,786	9
TOTAL	**29,026,031**		**5,861,886**			**14,296,786**	

State Deer-Hunting Regulations (A)

State	Is Gun Hunting From a Tree Stand Legal?	Is Blaze Orange Required?	Is Hunting Over Bait Legal?	Is Hunting Over Salt or Minerals Legal?	Is It Legal to Use Dogs to Trail Wounded Deer?	Is It Legal to Use Dogs to Hunt Deer?
Alabama	Yes	Yes	No	Yes	Yes	Yes
Arizona	Yes	No	No	Yes	No	No
Arkansas	Yes	Yes	Yes	Yes	Yes	Yes
California	Yes	No	No	No	Yes	Yes
Colorado	Yes	Yes	No	No	No	No
Connecticut	Yes	Yes	No	No	No	No
Delaware	Yes	Yes	No	No	No	No
Florida	Yes	Yes	Yes	Yes	Yes	Yes
Georgia	Yes	Yes	No	No	Yes	Yes
Idaho	Yes	No	No	No	No	No
Illinois	Yes	Yes	No	No	No	No
Indiana	Yes	Yes	No	No	No	No
Iowa	Yes	Yes	No	No	No	No
Kansas	Yes	Yes	Yes	Yes	No	No
Kentucky	Yes	Yes	Yes	Yes	No	No
Louisiana	Yes	Yes	Yes	Yes	Yes	Yes
Maine	Yes	Yes	No	No	No	No
Maryland	Yes	Yes	Yes	Yes	No	No
Massachusetts	Yes	Yes	No	No	No	No
Michigan	No	Yes	Yes	Yes	No	No
Minnesota	Yes	Yes	No	Yes	No	No
Mississippi	Yes	Yes	No	No	Yes	Yes
Missouri	Yes	Yes	No	Yes	No	No
Montana	Yes	Yes	No	No	No	No
Nebraska	Yes	Yes	Yes	Yes	Yes	Yes
Nevada	Yes	No	No	No	No	No
New Hampshire	Yes	No	Yes	No	No	No
New Jersey	Yes	Yes	Yes	Yes	No	No
New Mexico	No	No	No	No	No	No
New York	Yes	No	No	No	Yes	No
North Carolina	Yes	Yes	Yes	Yes	Yes	Yes
North Dakota	Yes	Yes	Yes	Yes	No	No
Ohio	Yes	Yes	Yes	Yes	No	No
Oklahoma	Yes	Yes	Yes	Yes	No	No
Oregon	Yes	No	Yes	Yes	No	No
Pennsylvania	Yes	Yes	No	No	No	No
Rhode Island	Yes	Yes	No	No	No	No
South Carolina	Yes	Yes	Yes	Yes	Yes	Yes
South Dakota	Yes	Yes	No	No	Yes	No
Tennessee	Yes	Yes	No	Yes	No	No
Texas	Yes	No	Yes	Yes	Yes	No
Utah	Yes	Yes	Yes	Yes	No	No
Vermont	Yes	No	Yes	No	No	No
Virginia	Yes	Yes	No	No	Yes	Yes
Washington	Yes	Yes	Yes	Yes	No	No
West Virginia	Yes	Yes	Yes	Yes	No	No
Wisconsin	Yes	Yes	Yes	Yes	No	No
Wyoming	Yes	Yes	Yes	Yes	No	No

State Deer-Hunting Regulations (B)

State	Minimum Hunting Age	Is Hunter Education Mandatory?	Is Bowhunting Education Mandatory?	Is a Separate Bowhunting-Education Class Offered?	Is Hunting From a Tree Stand Legal?
Alabama	None	Yes	No	No	Yes
Arizona	10	No	No	No	Yes
Arkansas	None	Yes	No	No	Yes
California	12	Yes	No	No	Yes
Colorado	14	Yes	No	Yes	Yes
Connecticut	12	Yes	Yes	Yes	Yes
Delaware	None	Yes	No	Yes	Yes
Florida	None	Yes	Yes	Yes	Yes
Georgia	None	Yes	No	No	Yes
Idaho	12	Yes	Yes	Yes	Yes
Illinois	None	Yes	No	No	Yes
Indiana	None	Yes	No	No	Yes
Iowa	None	Yes	No	Yes	Yes
Kansas	14	Yes	No	No	Yes
Kentucky	None	Yes	No	No	Yes
Louisiana	None	Yes	No	Yes	Yes
Maine	10	Yes	Yes	No	Yes
Maryland	None	Yes	No	No	Yes
Massachusetts	12	No	No	Yes	Yes
Michigan	12	Yes	No	No	Yes
Minnesota	12	Yes	No	Yes	Yes
Mississippi	None	Yes	No	No	Yes
Missouri	11	Yes	No	No	Yes
Montana	12	Yes	Yes	Yes	Yes
Nebraska	14	Yes	Yes	Yes	Yes
Nevada	12	Yes	No	No	Yes
New Hampshire	None	Yes	No	Yes	Yes
New Jersey	10	Yes	Yes	Yes	Yes
New Mexico	12	Yes	No	Yes	Yes
New York	14	Yes	Yes	Yes	Yes
North Carolina	None	Yes	No	No	Yes
North Dakota	14	Yes	Yes	Yes	Yes
Ohio	None	Yes	No	No	Yes
Oklahoma	None	Yes	No	No	Yes
Oregon	12	Yes	No	No	Yes
Pennsylvania	12	Yes	No	No	Yes
Rhode Island	12	Yes	Yes	Yes	Yes
South Carolina	None	Yes	No	No	Yes
South Dakota	12	Yes	Yes	Yes	Yes
Tennessee	None	Yes	No	Yes	Yes
Texas	None	Yes	No	No	Yes
Utah	14	Yes	No	No	Yes
Vermont	None	Yes	No	No	Yes
Virginia	None	Yes	No	No	Yes
Washington	None	Yes	No	No	Yes
West Virginia	None	Yes	No	No	Yes
Wisconsin	12	Yes	No	Yes	Yes
Wyoming	14	Yes	No	No	Yes

State Deer-Hunting Trends (A)

State	Deer-Vehicle Collisions	Deer Crop Damage	Number of Youth Hunters	Total Number of Hunters	Number of Male Hunters	Number of Female Hunters	Hunting With Rifles
Alabama	U	I	D	D	D	D	D
Arizona	C	C	U	I	I	I	I
Arkansas	U	C	U	U	U	U	C
California	U	U	U	D	U	U	C
Colorado	U	D	U	C	U	U	C
Connecticut	C	C	C	C	C	C	C
Delaware	I	I	U	I	I	I	NA
Florida	U	C	U	D	U	U	U
Georgia	I	C	U	C	C	I	C
Idaho	U	D	I	I	I	U	I
Illinois	I	U	U	I	U	U	U
Indiana	D	U	U	I	I	I	NA
Iowa	C	C	C	C	U	U	NA
Kansas	C	D	U	C	U	U	C
Kentucky	D	C	D	C	C	C	C
Louisiana	U	D	U	C	U	U	C
Maine	C	U	D	C	U	U	C
Maryland	C	C	D	C	U	U	C
Massachusetts	U	U	U	U	U	U	NA
Michigan	D	D	I	D	D	D	D
Minnesota	U	U	U	C	U	U	U
Mississippi	I	I	U	C	U	U	C
Missouri	D	D	U	I	U	U	C
Montana	U	D	U	C	U	U	U
Nebraska	C	D	U	C	U	U	C
Nevada	U	D	U	D	U	U	D
New Hampshire	D	C	C	C	U	U	C
New Jersey	C	C	D	D	U	U	NA
New Mexico	U	U	U	D	U	U	D
New York	D	C	I	D	U	U	U
North Carolina	C	C	C	C	C	C	I
North Dakota	I	I	I	I	U	U	I
Ohio	I	I	U	I	U	U	NA
Oklahoma	C	I	U	I	U	U	I
Oregon	U	U	D	D	U	U	D
Pennsylvania	I	D	I	D	U	U	D
Rhode Island	I	I	U	I	I	C	I
South Carolina	I	C	C	I	U	U	I
South Dakota	U	I	I	I	I	U	I
Tennessee	U	I	U	C	U	U	C
Texas	U	U	U	I	I	I	U
Utah	D	D	C	D	D	D	D
Vermont	I	I	U	D	U	U	D
Virginia	D	U	I	C	U	U	U
Washington	I	U	D	D	D	C	D
West Virginia	D	D	I	I	I	I	I
Wisconsin	C	C	U	C	U	U	C
Wyoming	D	U	I	D	D	D	D

C = Constant D = Decreased I = Increased NA = Not Allowed U = Unknown

State Deer-Hunting Trends (B)

State	Hunting With Shotguns	Hunting With Handguns	Hunting With Muzzle-Loaders	Hunting With Bows	Coyote Pressure on Deer Herd	Anti-Hunting Pressure	Problems With Poaching
Alabama	D	D	D	I	I	U	C
Arizona	I	I	I	C	C	C	U
Arkansas	C	C	C	C	U	U	U
California	C	C	C	I	U	C	U
Colorado	U	U	C	I	U	I	C
Connecticut	C	NA	C	I	C	C	C
Delaware	I	NA	I	I	U	I	C
Florida	U	U	D	D	C	I	C
Georgia	C	C	C	C	U	I	C
Idaho	U	U	I	I	U	C	U
Illinois	I	I	C	I	U	C	U
Indiana	I	D	I	I	I	U	U
Iowa	C	NA	C	C	U	U	U
Kansas	C	C	I	C	U	C	C
Kentucky	D	C	I	C	I	C	C
Louisiana	C	I	C	I	U	U	D
Maine	U	U	I	I	C	C	C
Maryland	C	C	I	I	U	I	C
Massachusetts	U	NA	I	I	U	C	U
Michigan	D	D	D	D	I	D	I
Minnesota	U	U	U	C	U	C	C
Mississippi	C	C	I	I	C	C	C
Missouri	C	C	I	I	U	C	U
Montana	U	U	U	U	I	C	C
Nebraska	U	U	I	C	U	C	U
Nevada	U	U	D	D	U	I	C
New Hampshire	C	C	I	I	I	C	I
New Jersey	D	NA	I	I	I	C	C
New Mexico	D	D	I	I	U	I	U
New York	U	U	I	I	C	C	U
North Carolina	D	I	I	C	U	C	C
North Dakota	I	I	I	I	U	I	C
Ohio	I	I	I	I	U	C	C
Oklahoma	C	C	C	C	I	C	C
Oregon	D	C	I	I	I	I	U
Pennsylvania	C	C	D	I	I	C	D
Rhode Island	C	NA	I	I	U	C	C
South Carolina	C	C	D	I	U	I	C
South Dakota	C	C	I	D	C	C	D
Tennessee	C	C	I	C	U	C	C
Texas	U	U	U	U	I	I	U
Utah	D	C	D	D	I	I	D
Vermont	U	U	I	I	C	I	C
Virginia	U	U	I	I	U	U	U
Washington	D	D	D	D	C	I	C
West Virginia	C	C	I	I	U	U	D
Wisconsin	C	I	I	I	C	I	C
Wyoming	C	C	C	C	D	I	C

C = Constant D = Decreased I = Increased NA = Not Allowed U = Unknown

Bowhunting Statistics

State	Season Dates	Season Days	Resident Hunters	Nonresident Hunters	Number of Deer Harvested	May Public Hunt With a Crossbow?
Alabama	10/15–1/31	108	62,000	6200	32,000	No
Arizona	8/20–1/31	74	18,262	621	758	Yes
Arkansas	10/1–2/28	151	Unknown	Unknown	14,797	Yes
California	7/9–9/11	65	31,870	Combined	1682	Yes
Colorado	8/28–11/26	30	14,300	9100	4900	Yes
Connecticut	9/15–12/31	108	13,000	2000	2165	No
Delaware	9/1–1/31	107	6725	402	507	Yes
Florida	9/11–11/14	75	30,613	515	Combined	Yes
Georgia	9/17–10/21	35	95,000	2000	21,450	No
Idaho	8/30–12/19	26	23,000	2734	3000	Yes
Illinois	10/1–1/12	104	81,000	Combined	21,000	No
Indiana	10/1–12/31	87	92,500	1500	23,933	No
Iowa	10/1–1/10	86	34,165	450	8814	No
Kansas	10/1–12/31	81	15,000	Just Started	5500	No
Kentucky	10/1–1/15	107	94,640	1800	8400	Yes
Louisiana	10/1–1/20	112	50,800	345	22,600	No
Maine	9/30–10/29	26	12,000	1175	682	No
Maryland	9/15–1/31	118	46,000	7000	11,043	No
Massachusetts	11/7–11/26	20	17,500	337	1370	No
Michigan	10/1–1/1	77	338,320	6200	99,990	No
Minnesota	9/18–12/31	104	70,000	1100	12,730	No
Mississippi	10/1–1/31	62	58,000	5587	41,000	No
Missouri	10/1–12/31	92	92,031	1544	14,696	No
Montana	9/3–10/16	44	16,900	8000	3473	Yes
Nebraska	9/15–12/31	99	13,177	621	3581	No
Nevada	8/14–1/2	42	1954	320	285	No
New Hampshire	9/15–12/15	91	17,804	3761	877	No
New Jersey	10/2–1/26	76	50,685	2002	17,006	No
New Mexico	9/1–1/15	96	5314	511	755	No
New York	9/27–12/31	96	160,000	5000	20,048	No
North Carolina	9/12–11/19	54	100,000	5385	10,000	No
North Dakota	9/3–12/31	121	11,400	669	4473	No
Ohio	10/2–1/31	122	192,000	2000	23,160	Yes
Oklahoma	10/1–12/31	78	53,106	403	7837	No
Oregon	8/27–12/4	53	26,555	1308	5391	No
Pennsylvania	10/2–1/8	49	29,247	22,857	49,407	No
Rhode Island	10/1–1/31	123	3800	200	378	No
South Carolina	8/15–10/10	41	25,000	5000	Combined	Yes
South Dakota	10/1–12/31	92	Combined	Combined	3400	No
Tennessee	9/25–11/14	44	82,040	5228	18,691	No
Texas	10/1–10/31	31	78,000	2000	19,500	No
Utah	8/21–9/17	28	22,332	1390	4704	No
Vermont	10/1–12/11	32	23,110	5592	2999	No
Virginia	10/1–1/7	43	63,273	2287	15,900	No
Washington	9/15–12/15	51	22,013	155	4856	No
West Virginia	10/15–12/31	67	107,160	18,047	26,425	No
Wisconsin	9/18–12/31	86	215,292	5535	52,623	No
Wyoming	9/1–9/30	30	8111	2669	1538	Yes
		TOTAL	**2,837,999**	**151,550**	**650,324**	

Muzzleloader-Hunting Statistics

State	Muzzle-Loader Season Dates	Special Season Days	Are There Special Muzzleloader Restrictions?	Resident Hunters	Nonresident Hunters	Deer Harvested
Alabama	11/20–1/31	0	Yes	23,000	2300	Combined
Arizona	10/29–12/31	32	Yes	1380	60	288
Arkansas	10/23–1/2	21	Yes	Unknown	Unknown	16,219
California	10/22–1/29	48	Yes	2071	Combined	86
Colorado	9/11–9/19	9	Yes	6500	1300	2500
Connecticut	12/12–12/24	12	Yes	6000	250	367
Delaware	10/10–1/25	9	Yes	7352	603	1045
Florida	10/15–2/27	20	Yes	14,089	237	Combined
Georgia	9/23–1/12	57	Yes	20,000	1212	Combined
Idaho	11/10–12/9	20	Yes	12,600	1498	6000
Illinois	12/9–12/11	3	Yes	3017	Combined	Combined
Indiana	11/13–12/19	16	Yes	45,400	225	13,621
Iowa	10/15–1/10	32	Yes	19,000	50	6564
Kansas	9/18–12/12	21	Yes	2575	Just Started	1100
Kentucky	10/15–12/16	9	No	35,000	1000	8200
Louisiana	12/6–12/10	5	Yes	4197	50	3200
Maine	11/29–12/4	6	No	5000	160	112
Maryland	10/20–12/31	16	Yes	34,717	5283	5174
Massachusetts	12/19–12/21	3	Yes	12,500	241	580
Michigan	12/3–12/19	17	Yes	172,250	4153	23,240
Minnesota	11/27–12/12	16	Yes	3000	58	1097
Mississippi	12/2–12/15	14	Yes	64,000	6165	39,000
Missouri	11/12–12/11	9	Yes	13,116	Just Started	2566
Montana	9/15–12/31	0	Yes	Combined	Combined	Combined
Nebraska	12/4–12/19	16	Yes	5819	127	2282
Nevada	9/11–1/2	0	Yes	1200	200	139
New Hampshire	10/30–11/9	11	No	20,608	2875	2369
New Jersey	12/13–1/1	14	Yes	20,000	436	5739
New Mexico	9/10–9/20	11	Yes	4727	822	1692
New York	10/15–12/20	14	Yes	43,000	900	3407
North Carolina	10/10–11/19	18	No	96,000	5169	13,000
North Dakota	11/26–12/6	11	Yes	646	0	299
Ohio	1/5–1/7	3	Yes	107,000	1000	10,396
Oklahoma	10/22–10/30	9	Yes	42,556	202	10,255
Oregon	10/1–12/11	72	Yes	Combined	Combined	868
Pennsylvania	12/27–1/8	12	Yes	70,941	5548	7606
Rhode Island	10/31–12/24	36	Yes	4500	250	624
South Carolina	10/1–10/10	10	Yes	Combined	Combined	Combined
South Dakota	10/27–12/19	42	Yes	Combined	Combined	300
Tennessee	10/16–12/12	30	Yes	94,662	5228	27,858
Texas	1/7–1/15	9	No	Combined	Combined	Combined
Utah	11/6–11/15	10	Yes	16,075	1020	3442
Vermont	12/3–12/11	9	Yes	13,288	1589	291
Virginia	11/7–1/7	29	Yes	58,617	3639	25,995
Washington	10/1–12/15	34	Yes	9161	69	2740
West Virginia	11/21–12/17	6	Yes	72,236	7378	9769
Wisconsin	11/29–12/5	7	Yes	2500	Combined	385
Wyoming	9/10–11/30	0	Yes	Combined	Combined	Combined
			TOTAL	**1,190,300**	**61,297**	**260,415**

Hunting Contacts

ALABAMA
Dept. of Conservation
and Natural Resources
64 N. Union St.
Montgomery, AL 36130
334/242-3486

ALASKA
Dept. of Fish and Game
P.O. Box 25526
Juneau, AK 99802
907/465-4100

ARIZONA
Game and Fish Dept.
2222 W. Greenway Rd.
Phoenix, AZ
85023-4312
602/942-3000

ARKANSAS
Game and Fish
Commission
#2 Natural
Resources Drive
Little Rock, AR 72205
501/223-6300

CALIFORNIA
Fish and Game
Commission
1416 9th St.
Rm. 1320
Sacramento, CA 95814
916/653-4899

COLORADO
Dept. of Natural
Resources
1313 Sherman
Rm. 718
Denver, CO 80203
303/866-3311

CONNECTICUT
Dept. of Environmental
Protection
79 Elm St.
Hartford, CT 06106-5127
203/424-3011

DELAWARE
Dept. of Natural
Resources
89 Kings Highway
Dover, DE 19903
302/739-5297

FLORIDA
Game and Fresh Water
Fish Commission
620 S. Meridian St.
Tallahassee, FL
32399-1600
904/488-1960

GEORGIA
Dept. of Natural
Resources Wildlife
Resources Division
2070 U.S. Hwy. 278, SE
Social Circle, GA 30279
770/918-6401

HAWAII
Dept. of Land and
Natural Resources
Division of Forestry
and Wildlife
1151 Punchbowl St.
Honolulu, HI 96813
808/587-0077

IDAHO
Fish and Game Dept.
600 S. Walnut St.
Boise, ID 83707
208/334-3700

ILLINOIS
Dept. of Natural Resources
524 S. Second St.
Rm. 400 LTP
Springfield, IL 62701-1787
217/785-0067

INDIANA
Dept. of Natural Resources
402 W. Washington St.
Rm. 160
Indianapolis, IN 46204
317/232-4200

IOWA
Dept. of Natural Resources
E. 9th and Grand Ave.
Wallace Building
Des Moines, IA 50319-0034
515/281-5145

KANSAS
Dept. of Wildlife and Parks
900 SW Jackson St.
Suite 502
Topeka, KS 66612-1233
913/296-2281

KENTUCKY
Dept. of Fish and Wildlife Resources
#1 Game Farm Rd.
Frankfort, KY 40601
502/564-3400

LOUISIANA
Dept. of Wildlife and Fisheries
P.O. Box 98000
Baton Rouge, LA 70898-9000
504/765-2800

MAINE
Dept. of Inland Fisheries and Wildlife
284 State St.
Station #41
Augusta, ME 04333
207/287-2766

MARYLAND
Dept. of Natural Resources
Tawes State Office Building
580 Taylor Ave.
Annapolis, MD 21401
410/974-3987

MASSACHUSETTS
Dept. of Fisheries, Wildlife and Environmental Law Enforcement
100 Cambridge St.
Rm. 1902
Boston, MA 02202
617/727-3155

MICHIGAN
Dept. of Natural Resources
Box 30444
Lansing, MI 48909
517/373-1263

MINNESOTA
Dept. of Natural Resources
500 Lafayette Rd.
St. Paul, MN 55155-4001
612/296-6157

MISSISSIPPI
Dept. of Wildlife, Fisheries and Parks
P.O. Box 451
Jackson, MS 39205
601/362-9212

MISSOURI
Dept. of Conservation
P.O. Box 180
Jefferson City, MO 65102-0180
573/751-4115

MONTANA
Dept. of Fish, Wildlife and Parks
1420 E. 6th St.
Helena, MT 59620
406/444-2535

NEBRASKA
Game and Parks Commission
2200 N. 33d St.
Lincoln, NE 68503-0370
402/471-0641

NEVADA
Dept. of Conservation and Natural Resources
Capitol Complex
123 W. Nye Ln.
Carson City, NV 89710
702/687-4360

NEW HAMPSHIRE
Fish and Game Dept.
2 Hazen Dr.
Concord, NH 03301
603/271-3422

NEW JERSEY
Division of Fish,
Game and Wildlife
501 E. State St.
CN 400,
Trenton, NJ 08625
609/292-2965

NEW MEXICO
Dept. of Game and Fish
P.O. Box 25112
Santa Fe, NM 87504
505/827-7911

NEW YORK
Dept. of Environmental
Conservation
50 Wolf Rd.
Albany, NY 12233
518/457-3730

NORTH CAROLINA
Wildlife Resources
Commission
Archdale Bldg.
512 N. Salisbury St.
Raleigh, NC 27604-1188
919/733-3391

NORTH DAKOTA
State Game
and Fish Dept.
100 N. Bismarck Expy.
Bismarck, ND 58501
701/328-6300

OHIO
Dept. of Natural
Resources
Fountain Square
Columbus, OH 43224
614/265-6565

OKLAHOMA
Dept. of Wildlife
Conservation
1801 N. Lincoln
Oklahoma City,
OK 73105
405/521-3851

OREGON
Fish and
Wildlife Division
400 Public Service Bldg.
Salem, OR 97310
503/378-3720

PENNSYLVANIA
Game Commission
2001 Elmerton Ave.
Harrisburg, PA
17110-9797
717/787-4250

RHODE ISLAND
Dept. of Environmental
Management
9 Hayes St.
Providence, RI 02903
401/277-2774

SOUTH CAROLINA
Dept. of
Natural Resources
Rembert C. Dennis Bldg.
P.O. Box 167
Columbia, SC 29202
803/734-3888

SOUTH DAKOTA
Game, Fish
and Parks Dept.
523 East Capitol
Pierre, SD 57501-3182
605/773-3387

TENNESSEE
Wildlife Resources
Agency
P.O. Box 40747
Ellington Agricultural
Center
Nashville, TN 37204
615/781-6500

TEXAS
Parks and Wildlife Dept.
4200 Smith School Rd.
Austin, TX 78744
512/389-4800

UTAH
State Dept. of Natural
Resources
1636 W. North Temple
Suite 316 , Salt Lake
City, UT 84116-3193
801/538-7200

VERMONT
Agency of Natural
Resources, Dept. of
Fish and Wildlife
103 S. Main St.
Waterbury, VT
05671-0501
802/241-3700

VIRGINIA
Dept. of Game and
Inland Fisheries
4010 W. Broad St.
Richmond, VA
23230-1528
804/367-1000

WASHINGTON
Dept. of Fish
and Wildlife
600 Capitol Way N.
Olympia, WA
98501-1091
360/902-2200

WEST VIRGINIA
Division of
Natural Resources
1900 Kanawha Blvd., E.
Charleston, WV 25305
304/558-2754

WISCONSIN
Dept. of Natural
Resources
Box 7921
Madison, WI 53707
608/266-2621

WYOMING
Game and Fish Dept.
5400 Bishop Blvd.
Cheyenne, WY
82006-0001
307/777-4600

Other Useful Addresses:

ARCHERY MANUFACTURERS ORGANIZATION
2622 C-4 NW 43d St.
Gainesville, FL 32606
904/377-8262

BOONE AND CROCKETT CLUB
Old Milwaukee Depot
250 Station Dr.
Missoula, MT 59801
406/542-1888

IZAAK WALTON LEAGUE OF AMERICA
707 Conservation Ln.
Gaithersburg, MD
20878
301/548-0150

NATIONAL MUZZLE-LOADING RIFLE ASSOCIATION
P.O. Box 67
Friendship, IN 47021
812/667-5131

NATIONAL RIFLE ASSOCIATION
11250 Waples Mill Rd.
Fairfax, VA 22030
703/267-1000

NATIONAL SHOOTING SPORTS FOUNDATION
Flintlock Ridge
Office Center
11 Mile Hill Rd.
Newtown, CT
06470-2359
203/426-1320

ORION—THE HUNTER'S INSTITUTE
P.O. Box 5088
Helena, MT 59604
406/449-2795

POPE AND YOUNG CLUB
P.O. Box 548
Chatfield, MN 55923
507/867-4144

SAFARI CLUB INTERNATIONAL
4800 W. Gates Pass Rd.
Tucson, AZ 85745
602/620-1220

UNITED CONSERVATION ALLIANCE
P.O. Box 820706
Houston, TX 77282
713/558-1399

U.S. FISH AND WILDLIFE SERVICE
Office of Public Affairs
1849 C St. NW
Rm. 3240
Washington, DC 20240
202/208-4131

U.S. GEOLOGICAL SURVEY
P.O. Box 25286
Denver, CO 80225
1-800/USA-MAPS

WHITETAILS UNLIMITED, INC.
P.O. Box 720
Sturgeon Bay, WI 54235
414/743-6777

WILDLIFE FOREVER
12301 Whitewater Dr.
Suite 210
Minnetonka, MN 55343
612/936-0605

THE WILDLIFE LEGISLATIVE FUND OF AMERICA
801 Kingsmill Pkwy.
Columbus, OH
43229-1137
614/888-4868

Successful hunters at a camp in Taylor County, Wisconsin, 1909.

Deer hunters at a cabin in Marinette County, Wisconsin.

Bowhunting legend Fred Bear with a buck he took in 1941.

Acknowledgments

This book could never have been put together without the guidance and assistance of Frank Golad, who has edited the *Sports Afield* Almanac since 1986, and has contributed to the magazine as an illustrator, writer and adviser since the early 1950s. He is the heart and soul of the Almanac, and for that reason he is the heart and soul of this book.

Almost all the photographs that accompany the main text were taken by Leonard Lee Rue III, who has photographed deer and other wildlife for more than 30 years. Photographs in the introductory section and at the end of the book have been provided by the Library of Congress (p. xii), the Vermont Historical Society (p. xiii), and the State Historical Society of Wisconsin (cover photo: serial number WHi [X3] 15441; pp. x–xi: WHi [X3] 1386; pp. 248–249: WHi [X3] 38817 [top], WHi [X3] 28589 [bottom]; and pp. 250–251 WHi [X3] 36232). Many of the illustrations that accompany the text were done by Glenn Wolff, except for those at the beginning of each chapter, which are Frank's.

It is not possible to thank all the writers who contributed to this book, but many of the pieces were written by our field editors: Anthony Acerrano, Gerald Almy, Grits Gresham, Tom Gresham, Ted Kerasote, Thomas McIntyre, Dwight Schuh and Contributing Editor Peter Fiduccia.

Thanks are also due to Carol Cammero, who typed most of the manuscript;
Sonje Berg, who was the copy editor;
and Terry McDonell, who thought the book was a good idea.